T0338444

PARALLEL ALGORITHMS AND ARCHITECTURES FOR DSP APPLICATIONS

THE KLUWER INTERNATIONAL SERIES
IN ENGINEERING AND COMPUTER SCIENCE

VLSI, COMPUTER ARCHITECTURE AND
DIGITAL SIGNAL PROCESSING
Consulting Editor
Jonathan Allen

Latest Titles

PARALLEL ALGORITHMS AND ARCHITECTURES FOR DSP APPLICATIONS

Edited by

Magdy A. Bayoumi
The University of Southwestern Louisiana

KLUWER ACADEMIC PUBLISHERS
Boston/Dordrecht/London

Distributors for North America:
Kluwer Academic Publishers
101 Philip Drive
Assinippi Park
Norwell, Massachusetts 02061 USA

Distributors for all other countries:
Kluwer Academic Publishers Group
Distribution Centre
Post Office Box 322
3300 AH Dordrecht, THE NETHERLANDS

Library of Congress Cataloging-in-Publication Data
Parallel algorithms and architectures for DSP applications / edited by
    Magdy A. Bayoumi.
        p.   cm. -- (The Kluwer international series in engineering and
    computer science ; SECS 149. VLSI, computer architecture, and
    digital signal processing)
        Includes bibliographical references and index.
        ISBN 0-7923-9209-4
        1. Signal processing--Digital techniques. 2. Parallel processing
    (Electronic computers) 3. Computer architectures. 4. Integrated
    circuits--Very large scale integration. I. Bayoumi, Magdy A.
    II. Series: Kluwer international series in engineering and computer
    science ; SECS 149. III. Series: Kluwer international series in
    engineering and computer science. kVLSI, computer architecture, and
    digital signal procession.
    TK5102.5.P352   1991
    621.3822--dc20   91-26492
                                                                     CIP

Printed on acid-free paper.

Printed in the United States of America

To the love of my life: Seham

Contents

PREFACE

Over the past few years, the demand for high speed Digital Signal Processing (DSP) has increased dramatically. New applications in real-time image processing, satellite communications, radar signal processing, pattern recognition, and real-time signal detection and estimation require major improvements at several levels; algorithmic, architectural, and implementation. These performance requirements can be achieved by employing parallel processing at all levels. Very Large Scale Integration (VLSI) technology supports and provides a good avenue for parallelism.

Parallelism offers efficient solutions to several problems which can arise in VLSI DSP architectures such as:

1. *Intermediate data communication and routing*: several DSP algorithms, such as FFT, involve excessive data routing and reordering. Parallelism is an efficient mechanism to minimize the silicon cost and speed up the processing time of the intermediate middle stages.

2. *Complex DSP applications*: the required computation is almost doubled. Parallelism will allow two similar channels processing at the same time. The communication between the two channels has to be minimized.

3. *Application specific systems*: this emerging approach should achieve real-time performance in a cost-effective way.

4. *Testability and fault tolerance*: reliability has become a required feature in most of DSP systems. To achieve such property, the involved time overhead is significant. Parallelism may be the solution to maintain acceptable speed performance.

Parallelism can be achieved at different levels; algorithms, architectures, and implementation. Most DSP algorithms have inherent parallelism in computation steps and data manipulation. The architecture band extends across different classes of parallelism; (1) using a set of Von Neuman processors and one or more shared memories, (2) achieving concurrency by employing an asynchronous timing paradigm, (3) large grain computation on a few powerful processors (such as Intel iPSC), (4) fine-grain computation on many simple processors (such as the Connection Machine), or (5) VLSI arrays which have regular and local communication, local control and I/O restricted boundaries. In achieving parallelism, several problems will arise which need to be tackled.

This book addresses several related issues and problems focussed on DSP algorithms and architectures. The main topics which have been discussed in this book are:

● *Communication*: This problem ranges from global but not very scalable solutions, like busses, through somewhat more scalable solutions like interconnection networks, to local but scalable point-to-point connection schemes. Several solutions have been offered which are DSP applications dependent; VLSI arrays for matrix-based computations (Chapter 6), mesh, pyramid, and combinations between them (Chapter 1), Orthogonal trees (Chapter 5).

● *Emerging Technologies*: Optical communication has been investigated with a case study of shuffle-exchange topology (Chapter 2). Neural Network (NN) technology provides a new paradigm for parallelism. Implementing NN on parallel architectures is an essential step to achieve the expected performance. Two case studies have been discussed; (1) Mapping multilayer perceptron NN onto SIMD arrays with achieving high throughput and flexibility (Chapter 9), and (2) mapping general NN onto fixed size array taking into consideration the used learning model (Chapter 10).

● *Design Environments and Experimentation*: With the increasing complexity of implemented systems, design environments, frameworks and simulation have become necessary design tools. Two examples have been given; one for asynchronous systems (Chapter 8) and another for shared memory architectures (Chapter 7). Analyzing the performance of an algorithm running on specific architectures can be used as guidelines for algorithm evaluation and implementation. As a case study, Fast Fourier Transform (FFT) algorithm has been analyzed using the Connection Machine and the PASM computer (a research computer at Purdue University) (Chapter 3).

● *Applications*: Three intensive computation case studies have been addressed; (1) Back-Projection Reconstruction (BPR) for Computer Tomography, (Chapter 5), (2) Adaptive Beamforming for Spatial Filtering >From an Array of Sensors (Chapter 4), and (3) Iterative Image Restoration (Chapter 1).

● *Algorithm Design*: The central point is to take advantage of the substantial parallelism of DSP algorithms and to obtain the maximum performance from complex parallel architectures. Parallelism on the algorithmic level can be employed for fault-tolerance (Chapter 4). Devising mapping methodologies for algorithms onto parallel architectures is gaining considerable interest; an example of mapping matrix-based computation on VLSI arrays is discussed (Chapter 6). These mapping methodologies are evaluated based on the implementation technology. In VLSI, AT^2 can be used as a performance measure (Chapter 1).

The intent of this book is to be informative and stimulating for the readers to gain knowledge and participate in fast evolving VLSI DSP field. It establishes a good understanding of the strength of parallel DSP in different applications and on various architectures. The book can be used as a textbook for research courses in VLSI, DSP, Parallel Processing, and DSP Architectures. It can be used as a supplementary text for graduate and senior undergraduate courses in VLSI Architecture and design for DSP applications. It can also serve as a material for tutorials and short courses in VLSI DSP Architectures, DSP systems design and Parallel Processing.

The idea of this book was motivated by a special session with the same title "Parallel Algorithms and Architectures for DSP Applications" at ISCAS 1990 in New Orleans. That session was sponsored by the VLSI Systems and Applications (VSA) Technical Committee of the Circuits and Systems Society. I extend thanks to the members of this committee and to the speakers of that session for supporting the idea of this book when it was in its infancy stage. My sincere appreciation to the VLSI Signal Processing Technical Committee of the ASSP society which provides a stimulating environment and a constructive infrastructure for VLSI Signal Processing activities. Special thanks to the authors who patiently spent considerable time and effort to have their research work reported in this book. It has been a stimulating and constructive experience working with such a group of highly motivated scholars. The environment in the Center for Advanced Computer Studies has been dynamic, inspiring and supportive for such project. My sincere thanks to Kluwer Academic Publishers for the enthusiasm they showed about this book, to Bob Holland, the editor and his assistant Rose Luongo for their support, encouragement, and patience. They have established a friendly communication channel for me.

Finally, I would like to acknowledge my lovely wife, *Seham*, and my interesting children; *Aiman*, *Walid*, and *Amanda* for their support and sacrifice during the course of this project. Seham does not believe that I finished my studies yet because of my working at night and during the weekends. I appreciate that she allows me to use our dining table as a desk because my office at home is ultra crowded. My younger son, Walid, always calls me Dr. Magdy Bayoumi, to remind my wife.

Magdy Bayoumi

PARALLEL ALGORITHMS AND ARCHITECTURES FOR DSP APPLICATIONS

1

PARALLEL ARCHITECTURES
FOR ITERATIVE IMAGE RESTORATION

M. Sarrafzadeh, A. K. Katsaggelos and S. P. R. Kumar

Department of Electrical Engineering
and Computer Science
Northwestern University
McCormick School of Engineering and Applied Sciences
Evanston, Illinois 60208-3118

Abstract

The recovery or restoration of an image that has been distorted is one of the most important problems in image processing applications. A number of algorithms or filters providing a solution to the image restoration problem, have appeared in the literature. Iterative restoration algorithms are used and analyzed in this work, due to advantages they offer over other existing techniques. Such algorithms, however, are generally computationally extensive and time consuming, as is the case with most image processing tasks. Therefore, there has been a natural interest in improving the response times of the image processors to extend the horizon of their applicability.

In this chapter parallel implementations of a class of iterative image restoration algorithms are proposed. More specifically, we propose Mesh, Pyramid, Mesh of Pyramids (MOP) and Pyramid of Meshes (POM) implementations of the iterative algorithms under consideration. MOPs and POMs are described as compositions of a Mesh and a Pyramid. Notions of network composition will be introduced. Area-time bounds on the proposed implementations are established. The efficiency of the proposed VLSI algorithms is evaluated by comparing the established bounds against lower bounds on AT^2, where A is the area of the VLSI chip and T is its computation time. The lower bounds for AT^2 which have been obtained for these architectures, explicitly indicate the dependence on the size of the filter support and the length of the operands. Often it is possible to alter the mathematical structure of the iteration to suit VLSI implementation, and gain efficiency in the restoration problem. This is illustrated via an example of a multi-step iteration for restoration.

1 INTRODUCTION

The recovery or restoration of an image that has been distorted is one of the most important problems in image processing applications [1]. A number of algorithms or filters providing a solution to the image restoration problem, have appeared in the literature [1]. Iterative restoration algorithms are used in this work, due to certain advantages they offer over other existing techniques. Among these advantages are the following [2, 3, 4] : i) there is no need to determine or implement the inverse of an operator; ii) knowledge about the solution can be incorporated into the restoration process; iii) the solution process can be monitored as it progresses; iv) constraints can be used to control the effect of noise. Iterative restoration tasks are generally computationally expensive and time consuming, as is the case with most image processing tasks. There has been a natural interest in improving the response times of the image processors to extend the horizon of their applicability. While early research in this direction focussed on exploiting the structure of the computation on a single processor (e.g., FFT algorithm), enhancing the speed by employing multiprocessors is currently of intense interest. Several image processing systems with multiprocessors, such as STARAN (a general purpose system employing an interconnection network [5]) have already been implemented with some success [6, 7]. The recent technological revolution, represented by very-large-scale integration (VLSI), has generated considerable interest in hardware implementation of complex operations (e.g., see [8, 9] for application in signal/picture processing).

In general, algorithm design is the development of better procedures to reduce the time to solve a given problem on a given computing system. Exploitation of a multiprocessor system requires a radical departure from the traditional Von Neumann environment. Detection of parallelism in sequential programs is essential to the discipline. The new challenge is to exploit properties of VLSI to build effective and efficient computing structures. The fundamental criteria of optimality are A, the area of the VLSI chip, and T, its computation time. The aim is to design architectures that use these two resourcees in an optimal manner.

In this paper we propose mesh, pyramid, and mesh of pyramids (MOP) VLSI implementations of an iterative image restoration algorithm. Notions of network composition will be introduced and MOP and POM will be described as "composition" of meshes and pyramids. The restoration process is in essence a two-dimensional (2D) deconvolution process. An iterative algorithm with first-order or linear rate of convergence is considered in detail; it performs the deconvolution iteratively, carrying out a 2D convolution in each step, eventually converging to the original image.

The 2D convolution algorithms that have been proposed in the literature [8, 10, 11, 12] are not attractive in this context, since the image has to be stored and convolved with a mask (in-place image) repeatedly till convergence. This fact is kept in mind in studying the implementation here. In addition, VLSI layout and area-time complexity of the implementations are presented, along with lower bound analysis. While mesh is an attractive implementation due to regularity, the mesh of pyramids is shown to yield the fastest circuit.

This chapter is organized in the following manner. In Sec. 2 the form and the properties of the first-order iterative algorithm are described. The VLSI implementations considered in this work, are presented in Sec. 3. In Sec. 4 a multistep iteration is introduced. Finally, in Sec. 5 conclusions and current research directions are described.

2   ITERATIVE RESTORATION ALGORITHMS

An appropriate mathematical model of the image distorting process is the following [1]

$$y(i,j) = d(i,j) \star \star x(i,j) + v(i,j) \qquad (2.1)$$

where $y(i,j), x(i,j)$ and $v(i,j)$ are respectively the distorted and original images and the noise, $d(i,j)$ is the impulse response of the distortion system and $\star\star$ denotes 2D convolution. We assume, without lack of generality, that the original and distorted images are of the same size. By stacking or lexicographically ordering $M \times N$ images into MN vectors, Eq. (2.1) takes the following matrix form.

$$y = Dx + v, \qquad (2.2)$$

where D is a block-Toeplitz square matrix. It is mentioned here that Eq. (2.2) represents a more general model, that is, when the degradation is space-variant, in which case D has no particular structure. Although in the work presented in this paper the model of Eq. (2.1) is used, our results can be extended to include the more general case of space-variant degradations in a straightforward way. The *image restoration problem is then to invert Eq. (2.2) or to find an image as close as possible to the original one subject to a suitable optimality criterion, given y and D.*

A number of approaches can be found in the literature for solving the image restoration problem [1]. In this work we follow the regularization approach presented in [3, 4]. Such an approach results in obtaining a restored image by solving the following set of linear equations

$$(D^T D + \alpha C^T C)x = D^T y, \tag{2.3}$$

$$Ax = g, \tag{2.4}$$

where T denotes the transpose of a vector or matrix and α, the *regularization parameter* is inversely proportional to the signal to noise ratio (SNR). The matrix C represents a high-pass filter, such as the 2D Laplacian operator, which is chosen in such a way so that the energy of the restored image at high frequencies (due primarily to the noise amplification) is bounded [3, 4].

Equation (2.4) may be solved through a variety of numerical techniques. A successive approximations iterative algorithm is used in this work for restoring noisy-blurred images, due to its advantages, as was mentioned in the introduction [2, 3]. If Eq. (2.4) has one or more then the minimum norm solution can be successively approximated for $0 < \beta < 2||A||^{-1}$, by means of the following iteration [2, 3]

$$
\begin{aligned}
x_0 &= \beta g \\
x_{k+1} &= (I - \beta A)x_k + \beta g. \tag{2.5}
\end{aligned}
$$

If Eq. (2.4) does not have a solution, then the following iteration

$$
\begin{aligned}
x_0 &= \beta A^T g \\
x_{k+1} &= (I - \beta A^T A)x_k + \beta A^T g \\
&= W x_k + f, \tag{2.6}
\end{aligned}
$$

Where $W = I - \beta A^T A$ and $f = \beta A^T g$, converges to its minimum norm least squares solution x^+, defined by $x^+ = A^+ g$, where A^+ is the generalized inverse of A, for $0 < \beta < 2||A||^{-2}$. Algorithms (2.5) and (2.6) exhibit linear rate of convergence, since it can be shown that [13]

$$\frac{||x_k - x^+||}{||x^+||} \leq c^{k+1}, \tag{2.7}$$

where

$$c = max\{|1 - \beta||A||^2|, |1 - \beta||A^+||^{-2}|\}. \tag{2.8}$$

Iterations with higher convergence rates are also studied in [13]. We observe that iterations (2.5) and (2.6) have the same computational form. Therefore, without lack of generality, in the following, we will concentrate on iteration (2.6). The pointwise or unstaked version of iteration (2.6) is useful in considering different ways in implementing it in VLSI. When C in Eq. (2.3) models a space invariant constraint system with impulse response $c(i,j)$ (matrix C is approximated by a block-circulant matrix), A in Eq. (2.4) is also a block-circulant matrix and it is characterized by the impulse response $a(i,j) = d(-i,-j) \star \star d(i,j) + \alpha c(-i,-j) \star \star c(i,j)$. Then, the pointwise version of iteration (2.6) is given by

$$
\begin{aligned}
x_0(i,j) &= \beta a(-i,-j) \star \star g(i,j) \\
x_{k+1}(i,j) &= x_k(i,j) + \beta a(-i,-j) \star \star [g(i,j) - a(i,j) \star \star x_k(i,j)] \\
&= [\delta(i,j) - \beta a(-i,-j) \star \star a(i,j)] \star \star x_k(i,j) \\
&+ \beta a(-i,-j) \star \star g(i,j) \\
&= w(i,j) \star \star x_k(i,j) + f(i,j) \tag{2.9}
\end{aligned}
$$

where $w(i,j) = \delta(i,j) - \beta a(-i,-j) \star \star a(i,j)$, $f(i,j) = \beta a(-i,-j) \star \star g(i,j)$ and $\delta(i,j)$ is the 2D impulse function. Clearly, the pointwise version of iteration (2.5) is obtained from iteration (2.9) by omitting the convolution with $a(-i,-j)$.

A priori knowledge about the solution can be incorporated into the algorithm with the use of constraints [2]. Such a constraint can be represented by a projection operator which projects a signal onto a convex set of signals with certain a priori known properties [14]. An example of such a property is the positivity property, according to which each entry of the vector x is a nonnegative number since it represents light intensity. Then at each iteration the signal x_k is projected onto one or more convex sets before it is used in generating the next estimate of the restored image x_{k+1}. When the projection operator represents a pointwise or a local neighborhood based operation, it can be incorporated into the structure of the processor (to be presented in Sec. 3) in a straightforward way. For ease of exposition we will assume in the following that the projection operator is the identity.

3 VLSI IMPLEMENTATIONS

In this section, we will investigate the VLSI complexity of the iterative image restoration algorithm of Eq. (2.9). Note that at each step of iteration (2.9) a 2D convolution must be performed. However, since a 2D convolution is required at each iteration step we must store the entire image in the chip (informally, we have to "pay" for storing the image $x_k(i, j)$). Thus, we cannot employ previous two-dimensional convolution algorithms, since the I/O requirements would severe the performance of the system (see section 3.2 below). We shall refer to the two-dimensional convolution of interest, where the entire image must be stored in the chip, as *static two-dimensional convolution (S2DC)*. It should be kept in mind that the overall objective of the algorithms to be implemented is not convolution, but deconvolution instead, by means of a series of convolutions.

First, we will briefly review the VLSI model of computation. Then a lower bound on area-time measure of S2DC will be derived. Finally, we propose mesh, pyramid, and mesh of pyramid implementations of S2DC (see [15, 16, 17] for related results).

3.1 VLSI Model of Computation

In this section, first we review the VLSI model of computation and discuss computational limits of VLSI. We will implement our image restoration algorithm on a *mesh, pyramid,* and *mesh of pyramid.* Meshes and pyramids have been proven effective for a number of problems in digital signal processing. However, their combination has not been studied (also, VLSI complexity of pyramid has not been investigated).

We briefly review the synchronous model of VLSI computation [18, 19, 20]. A computation problem Π is a Boolean mapping from a set of input variables to a set of output variables. The mapping embodied by Π is realized by a Boolean machine described as a computation graph, $G = (V, E)$, whose vertices V are information processing devices or input/output ports and whose edges E are wires. A VLSI chip is a two-dimensional embedding of this computation graph according to the prescriptions of the model. The model is characterized by a collection of rules concerning layout, timing, and input/output (I/O) protocol: in addition, the model restricts the class of computation graphs to those having bounded fan-in and fan-out.

The layout rules are:

1. Wires (edges) have minimum width λ and at most ν wires ($\nu \geq 2$) can overlap at any point.

2. Nodes have minimum area $c\lambda^2$, for some $c \geq 1$.

No loss of generality is incurred if the layout is restricted to be an embedding of the computation graph in a uniform grid, typically the square grid: the latter is the plane grid, the vertices of which have integer coordinates (*layout grid*).

The timing rules specify that both gate switching and wire propagation of a bit take a fixed time τ_0 (hereafter, assumed equal to 1; see [21] for validity of the unit-delay model), irrespective of wire length (*synchronous* system). In addition, the I/O protocol is *semellective* (each input is received exactly once), *unilocal* (each input is received at exactly one input port), and *time-* and *place-determinate* (each I/O variable is available in a prespecified sequence at a prespecified port, for all instances of the problem). Two other types of I/O protocol constraints are normally considered: the *word-local* assumption and the *word-serial* assumption. An I/O protocol is *word-local* if, for any cut partitioning the chip, $o(s)$ input (output) words have some bit entering (exiting) the chip on each side of the cut [22], where s is the input size. This constraint is used in the derivation of the AT^2 lower bound and is adhered to in the construction of the upper bounds (designs). An I/O protocol is *word-serial* if, at any time instant, $o(s)$ input (output) words have some, but not all, of their bits read (written). This constraint is used in the derivation of the A lower bound and is adhered to in the construction of the minimal area circuit.

3.2 Lower Bound

Thompson [18, 23] established a now widely-used technique [20, 21, 22, 23] for obtaining area-time lower bounds by quantifying the *information exchange* required to solve the problem Π. (Also, see [24] for a generalized approach). This quantity, denoted by I, is defined as the minimum number of bits that two processors must exchange in order to solve Π when exactly half of the input variables of Π are available to each processor at the beginning of the computation.

More formally, consider a problem $\Pi(s)$, where s is the input size and a chip C_Π with area A that is capable of solving Π in time T. Consider a cut that partitions C_Π into the left side L and the right side R, such that each side reads about half of the input (i.e., $s/2 - o(s)$), as shown in Fig. 1a. The two processors, P_L and P_R, associated respectively with L and R cooperate to solve $\Pi(s)$ (see Fig. 1b). We denote by $I(s)$ the number of bits that P_L and P_R communicate to solve $\Pi(s)$. Clearly, $I(s)$ depends on the distribution of input/output bits between P_L and P_R, and this, in turn, depends on input/output protocol of C_Π.

The history of the computation performed by C_Π can be modeled with an area-time solid, as shown in Fig. 1c. The communication channel between P_L and P_R is represented by rectangle F (dashed line) that transects the longer of the two area dimensions. Thus, F has sides of length T and (at most) $\sqrt{A}$. So A_F, the area of F, is at most $\sqrt{A}T$. If $I(s)$ bits must flow across this channel then $A_F = \Omega(I(s))$. Hence, we obtain

$$\sqrt{A}T = \Omega(I(s)). \qquad (3.1)$$

With a suitable change in I/O protocol semantics [22], information exchange arguments also give lower bounds on area, namely, $A = \Omega(I)$ [12].

Consider an instance of a generalized two-dimensional convolution given by any two matrices $A_{N\times N}$ and $B_{(2P+1\circ[249z])\times(2P+1)}$, where each element is represented by $O(b)$ bits. As shown in [12], any VLSI chip that computes a two-dimensional convolution of A and B must satisfy $I = \Omega(NPb)$. Thus, $AT^2 = \Omega(N^2P^2b^2), A = \Omega(NPb)$, and (due to bounded fan-in) $T = \Omega(log(Nb))$, assuming $N \geq P$.

In static two-dimensional convolution (S2DC) $A = \Omega(N^2b)$, by definition, and $T = \Omega(log Pb)$. Clearly, $AT^2 = \Omega(N^2P^2b^2)$, for this bound has been established for an arbitrary two-dimensional convolution. *Note that due to area constraints we cannot employ previous designs with area $A = o(N^2b)$* (e.g., designs proposed in [10, 11, 12]).

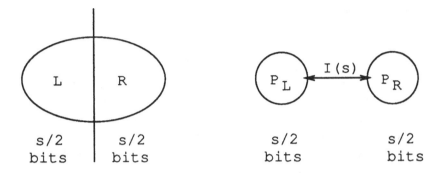

a. Bisection of a chip   b. Two-processor system

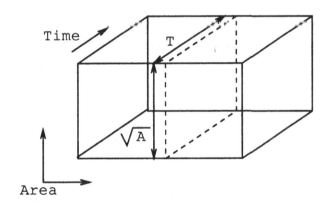

c. Area-time solid

Figure 1: AT^2 lower bounds

3.3 Mesh Implementation

Before we explain the architecture, the following point is worth noting. Each step of the restoration algorithm (Eqs. 2.6 or 2.9) involves convolution or matrix vector multiplication. The systolic implementation proposed in [[8], Ch. 12] for 2-D convolution converts the matrix into a linear array, and does not appear suitable for image restoration considered here.

The mesh implementation falls into the class of array processor architectures that have been extensively proposed and implemented for image processing tasks. See, for example, the cellular array machines discussed in [8]. In recent years, very large array processor systems have been built and reported. These include the 96×96 cellular logic image processor, the 128×128 massively parallel processor, and the connection machine which has about 100,000 processing cells. In each of the above implementations, each processor is substantially more powerful than a processor in the restoration implementation described below. The processors are organized as a two dimensional array. For convenience we assume that there is one processor per pixel. Figure 2a, depicts such a two-dimensional array, with one processor per pixel. A 256×256 picture requires 64K processors which do elementary computations that will be explained later. A smaller number of processors, say one for each 4×4 square in the picture requiring 4K processors, may be chosen. In this case each processor will be more complex. The architecture described for the case of one processor per pixel can be easily extended to the case of fewer processors. On the other hand, a fixed aray of processors can be used in restoring an image of any size. This can be done by partitioning the available image into subimages and restoring each subimage separately. The overlap-save or overlap-add block convolution technique needs to be implemented in this case in order to avoid errors at the boundaries of each subimage [25].

Here, we shall analyze VLSI complexity of the proposed implementations in the *word model*, where each word consists of one bit. Thereafter, we will show how to generalize our design to *the bit model*, where each word consists of b bits.

An implementation of iteration (2.9) (the pointwise version of Eq. (2.6)) is described in the following. We assume for simplicity that the impulse response $w(i,j)$ in Eq. (2.9) has support $(2P + 1) \times (2P + 1) = Q$ pixels. Each processor (i,j) corresponding to pixel (i,j) has

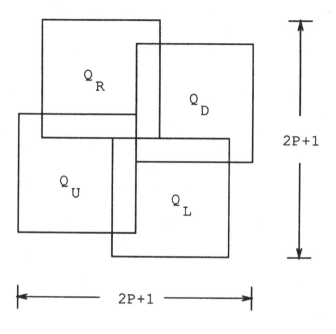

Figure 2: Partition $M_x^{(i,j)}$

a register to hold $f(i,j)$, and two other sets of registers to hold the weights $\omega(l,m)$ for $-P \le l, m \le P$, and all the restored image values in the $(2P+1) \times (2P+1)$ neighborhood of (i,j). The latter set of registers is denoted by $M_x^{(i,j)}$. The weights and $f(i,j)$, once loaded into the processors, remain unchanged throughout the course of the restoration computation. The contents of $M_x^{(i,j)}$, however, change over time in a manner to be described later. For the processors located P pixels or less away from the boundaries of the image, certain of the contents of $M_x^{(i,j)}$ will be fixed, representing the boundary conditions necessary for each convolution at each iteration step. These boundary values are usually set to be equal to zero but other scenarios can be also considered (the circular convolution scenario, for example.)

Each step of the restoration algorithm can be implemented in two phases: a communication phase followed by a computation phase. In the communication phase, each processor sends and receives messages, and thereby gathers the relevant partial results of the restored pixel values (i.e., the values of all neighbors with distance P or less) necesssary for the convolution in the computation phase. The obvious redundancies present in processor (i, j) in transmitting the entire matrix $M_x^{(i,j)}$ to all four of its neighbors, can be removed in several ways. One way is to partition $M_x^{(i,j)}$ into four parts $(Q_R, Q_D, Q_L,$ and $Q_U)$, as shown in Fig. 2. Each partition represents a quadrant of $M_x^{(i,j)}$ with outermost row or column omitted. For example, the partition Q_R of processor (i, j) is the set of register values $M_{x(l,m)}^{(i,j)}$ where $-P < l \leq 0$ and $0 \leq m \leq P$. Each partition has $P(P+1)$ values. It is easy to see that, in the communication phase, each processor needs to send only one partition to each processor. Specifically, the partitions Q_R, Q_L, Q_D and Q_U are respectively transmitted to the right, left, down, and up neighbors. These transmissions in the communication phase can be completely word-serial. This would require $2P(2P + 1)$ communication steps in each communication phase, and number of wires connecting neighboring processors need only be $O(1)$. Alternatively, the transmissions can be done on a part word-serial, part word-parallel basis as follows.

Communication phase for processor (i, j)

- step 0: send $x(i, j)$

For $1 \leq i \leq 2P - 1$, in

- step i :

    1. Copy messages received in step $(i - 1)$ to relevant locations in $M_x^{(i,j)}$ at distance i from the center.
    2. Send the values at distance i from the center, in partition Q_m to the m-neighbor, where $m\epsilon\{R, L, D, U\}$.

In communication step i of the above process, the number of words transmitted in parallel equals $\max\{(i + 1), 2P\}$, if $i \leq P$ and equals $2P - i$, if $P < i \leq 2P - 1$. This process would require $O(P)$ wires for connecting neighboring processors, and the total time for a communication phase is $O(P)$. If transmissions are completely word-parallel, the wire-width required is $O(P^2)$, which would be higher than the sides of the processor, and this would be inefficient. Following each communication phase, each processor performs the convolution computation (Eq. (2.9)) locally.

In the above implementation, each of the N^2 proceessors has $(2Q+1)$ registers, Q multipliers, and $O(logQ)$ adders. The width of the interprocessor connections is $O(P)$. Having $O(P)$ connections between two adjacent processors, allows them to communicate $O(P)$ words in $O(1)$ time. Thus, assuming $O(1)$ area for each register, multiplier and adder, the total area required is $O(NP) \times O(NP)$, that is, $O(N^2P^2)$. It takes $O(P)$ time for each communication phase, $O(1)$ time to multiply, and $O(log P)$ time to add. Thus, $AT^2 = O(N^2P^4) = O(N^2Q^2)$. We employ an optimal bit-multiplier for multiplying 2 b-bit numbers [26, 27], with $A_1 = b^2/T_1^2$ for $T_1 \epsilon [O(log b), O(\sqrt{b})]$. Essentially we place a bit-optimal multiplier where we had a unit multiplier in the word-model. Thus, the area of the unit multiplier is multiplied by A_1 and its time by T_1. We conclude:

Theorem 1: A mesh implementation of S2DC works in $O(PT_1^2)$ time and has area $O(N^2P^2b^2/T_1^2)$ for $T_1 \epsilon [O(log b), O(\sqrt{b})]$.

We can modify the above implementation in several ways. For example, we can make the width of the interprocessor connection $O(1)$. Doing so, reduces the area. However, the time will be increased. In that manner, the interprocessor connection can be set to any value X between 1 and P. Selecting different values of X offers a tradeoff between area and time.

3.4   Pyramid Implementation

A *pyramid* architecture $\Phi(N)$ consists of $log_4 N^4$ levels. At level 1 there is an $N \times N$ array of processors interconnected as a mesh. At level i, $2 \leq i \leq log_4 N^4$, there are $N^2/4^{i-1}$ processors interconnected as a mesh; each processor is connected to 4 processors at level $i-1$ (see Fig. 3a), as described in [28, 29]. An algorithm on a pyramid for solving generalized two-dimensional convolution, and thus for solving $S2DC$, has been proposed [29]. However, VLSI complexity thereof (e.g., layout issues, bit complexity) has not been analyzed.

We briefly outline the S2DC algorithm of Chang et. al. (for details, see [29]). Let X denote the $N \times N$ matrix with elements $X(i, j) = x(i, j)$, and W the $(2P+1) \times (2P+1)$ matrix with elements $W(i, j) = w(i + P + 1, j + P + 1)$.

procedure Pyramid-S2DC(X,W);
   begin $\Phi(N)$: a pyramid containing X and $(N/(2P+1))^2$
             copies of W;
    for $i = 1$ to Q *pardo* ($\star$ $Q = (2P+1)(2P+1)$ $\star$)

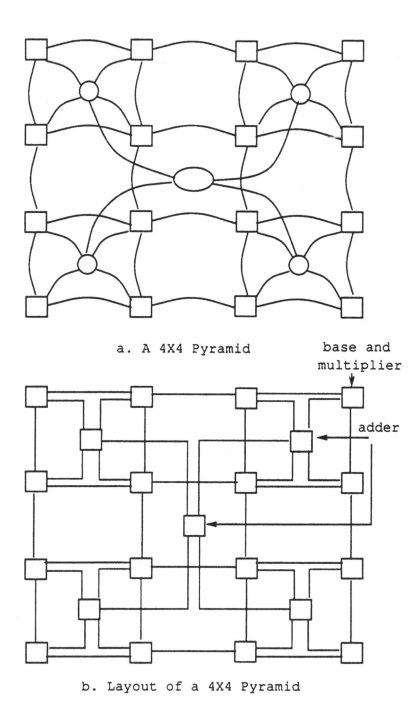

a. A 4X4 Pyramid

base and
multiplier

adder

b. Layout of a 4X4 Pyramid

Figure 3: A pyramid structure

begin calculate $X(a, b)$ using the third dimension;
    shift X in $P(N)$
  end
end.

Calculation of each $X(a, b)$ involves multiplying, element-by-element, two arrays of size $(2P + 1) \times (2P + 1)$. This task can be accomplished in $O(1)$ time, in parallel. The elements of the resulting array are added using the "third dimension", that is, using the pyramid. These numbers can be added in $O(log P)$ time in the word model. There are P^2 elements that use the same sub-pyramid. Thus, as soon as one set of element-by-element computation is finished, the next set will begin. Pipelining the computation gives a total of $O(P^2 + log P)$ time. Thus, $T = O(P^2)$.

Next, we shall focus on layout of a pyramid. At each level of $\Phi(N)$, processors are interconnected as a mesh and each processor is connected to four processors at the previous level (see Fig. 3b).

Each processor at level 1 contains a multiplier. Processors at level 2 to level $log(2P + 1)$ contain an adder and a broadcaster. (Hereafter, we shall use $loga$ to mean $log_4 a$). Numbers are multiplied at level 1, added at level 2 to level $log(2P + 1)$, and broadcasted from level $log(2P + 1)$ back to level 1. In the word model, each multiplier and each adder requires $O(1)$ area and operates in $O(1)$ time. Thus level i has height $H_i = H_{i-1}/2$, and since $H_1 = 2N$, then the total height is $O(N)$. Similarly, the total width is $O(N)$, or $A = O(N^2)$. We conclude that $AT^2 = O(N^2 P^4)$, in the word model. Now, assume that each element of X and W, is represented by b bits. Multipliers with $AT^2 = O(b^2)$, for multiplying two b-bit numbers are known [26, 27]. More precisely, $A_1 = O(b^2/T_1^2)$ for $T_1 \epsilon [O(log b), O(\sqrt{b})]$. A serial adder, with area $O(1)$ can be placed in each processor at level 2 to level $log P$. We conclude:

Theorem 2: A pyramid implementation of S2DC works in $O(P^2 T_1^2)$ time and has area $O(N^2 b^2/T_1^2)$ for $T_1 \epsilon [O(log b), O(\sqrt{b})]$.

We can insert more pyramids in the network to reduce the computation time while increasing the area. Indeed, we may start with a mesh (described in subsection 3.3) and add pyramids (regularly) to it. The more pyramids are added the faster is the computation and the larger the area. Therefore, a "natural" tradeoff between area and time is introduced. The other extreme instance of the just described tradeoff (mesh architecture being one of the extreme) is described in the next subsection.

3.5 Fastest Circuit

In practice it is often desired to restore an image as fast as possible. Here, we will focus on designing a fastest circuit for static two-dimensional convolution. As discussed earlier, the fastest circuit for S2DC operates in $O(log\,Pb)$ time (due to bounded fan-in).

Consider an $N \times N$ image X stored in an $N \times N$ mesh. Let B_i denote an arbitrary $(2P+1) \times (2P+1)$ block of X. Each element of B_i (for all i) is to be multiplied by an element of the coefficient matrix W, that is, to obtain $B_i(a,b)W(a,b)$, $1 \leq a,b \leq (2P+1)$. Finally, we must form the sum $s_i = \sum_{a,b} B_i(a,b)W(a,b)$. The sum s_i can be obtained in the following manner. At a processor containing $B_i(a,b)$ we store $W(a,b)$ and also place a multiplier. On the set of processors defined by B_i we place a pyramid $\Phi_i(2P+1)$ with $log(2P+1)$ levels. Each processor at level 2 to level $log(2P+1)$ contains an adder. In $O(1)$ time, in the word model, we can form $B_i(a,b)W(a,b)$, $1 \leq a,b \leq (2P+1)$. In $log(2P+1)$ time, we can form s_i, in the word model. As described in subsection 3.4, s_i can be obtained in $O(\log(2P+1)b)$ time in the bit model.

If we place a pyramid $\Phi_i(2P+1)$ on each block B_i then all sums can be obtained, in parallel, in $O(log(2P+1)b)$ time (assuming each processor contains the entire coefficient matrix). Thus, a fastest circuit is obtained, as shown in Fig. 4. We shall focus on area complexity of the proposed circuit.

First, we will consider the word model. Each base processor stores W and thus requires $O((2P+1)^2)$ area. Each pyramid $\Phi_i(2P+1)$, as described in subsection 3.4, contributes $O(2P+1)$ lines to the width and $O(2P+1)$ lines to the height. Consider an arbitrary vertical column of the base mesh. There are (at most) $N(2P+1)$ pyramids using one of the processors of this column. Each pyramid contributes $O(2P+1)$ lines to the height (see subsection 3.4). Thus, all pyramids, collectively, contribute $O(N(2P+1)^2)$ lines to the height. Similarly, their contribution to the width is $O(N(2P+1)^2)$. As described in subsection 3.4, multipliers and adders can be laid out within the same area-time.

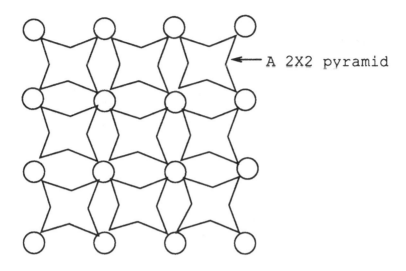

A 2X2 pyramid

Figure 4: A fastest circuit

Lemma 1: A mesh of pyramids implementation of S2DC works in $O(\log Pb)$ time and has area $O(N^2 P^4)$.

A modification of the proposed implementation yields improvement on the area bound. Such a modification results by assigning one processor to each $\sqrt{\log(2P+1)} \times \sqrt{\log(2P+1)}$ region of the image and letting each such block use the same pyramid. Thus the area is reduced by a factor of $O(\log P)$ and the time is increased by an additive term of $O(\log P)$. That is, we obtain:

Theorem 3: A mesh of pyramids implementation of S2DC works in $O(\log Pb)$ time and has area $O(N^2 P^4 / \log Pb)$.

The just described architecture, although quite fast, is rather complex. In the current VLSI technology, the implementation of such a system, is not feasible, especially for large images. However, the proposed implementation demonstrates what can be done with the future technology. It also provides new insights into the complexity of S2DC. Depending on the application and the technology, one of the proposed family of architectures can be used.

Note that the AT^2 bound of the mesh and pyramid is better than that of the mesh of pyramids by a factor of $O(\log Pb)$. Although the

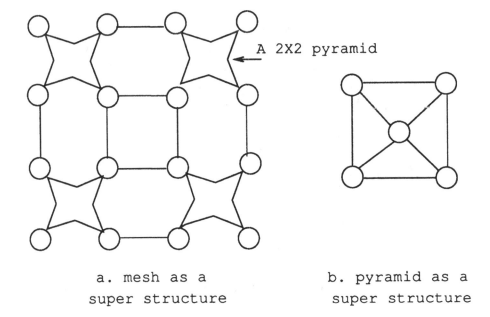

A 2X2 pyramid

a. mesh as a
super structure

b. pyramid as a
super structure

Figure 5: Network composition

AT^2 bounds of the mesh and pyramid are asymptotically equivalent, the mesh AT^2 bound has better constants. Furthermore, the maximum wire length in the mesh architecture is $O(1)$, while the maximum wire length in the pyramid and MOP architectures is $O(P)$.

For all three designs the I/O time is $O(N)$. Note that pyramid based architectures are more suitable for our problem than hypercube or mesh-of-trees architectures, since we only need "short-distance" communications. That is, only processors at distance $O(P)$ or less need to communicate.

3.6 Composition of Networks

In subsection 3.5, we constructed a network that involved both the mesh architecture and the pyramid. Here, we give a formal description of such "involvement' See [30] for a detailed discussion. We consider one-dimensional networks, that is networks that can be specified completely by one parameter. Square mesh and pyramid are both one-dimensinal networks. However, our discussion can be readily extended to higher dimension networks.

Consider two (one-dimensional) networks $N_1(n_1)$ and $N_1(n_2)$, where n_1 and n_2 are the size of N_1 and N_2 respectively. A composition of N_1 and N_2 is denoted by a four-tuple $\eta = [N_1(n_1), N_2(n_2), P_1, P_2]$. Network η consists of N_1 as a super structure and a set of N_2's. Copies of N_2 are placed in N_1 at every P_1 units in one direction and P_2 units in the other direction.

Figure 5a shows a 4×4 mesh $M(4)$ as a super structure. A collection of 2×2 pyramids $P(2)$ are repeated every 2 horizontal and every 2 vertical units. The same networks with the pyramid repeated every 1 unit is shown in Fig. 4. Fig. 5b contains a 2×2 pyramid as a super structure collection of 2×2 meshes every ∞ units (i.e., there is only one such mesh). Equivalently we could have denoted the network by $[P(2), M(2), 2, 2]$ or $[M(2), P(2), \infty, \infty]$.

The super structure tells us what kind of "global" communication is being performed and the secondary structure reveals the type of required "local" communication. Since in our S2DC problem, only "local" communication is needed (i.e., at unit $2P + 1$ element arrays) then it is appropriate to use a mesh as the primary structure and the pyramid on the secondary structure. Networks with pyramid on the primary structure and mesh on the secondary structure, that we call pyramid of meshes, are not suitable for our problem.

General properties of network composition is currently under investigation.

4 A MULTI-STEP ITERATION AND IMPLEMENTATION

In the previous section, we analyzed various VLSI implementations of an iterative restoration algorithm presented in Sec. 2. Often, it may be advantageous to alter the mathematical structure of the iteration itself to gain effectively in the restoration process. The algorithm of Sec. 2 is a single-step iteration in the sense that the (partially restored) image values at the current step depend only on the image values at the previous step. In this section, we discuss a multi-step iteration algorithm, which has a different mathematical structure and convergence behaviour. Nevertheless, it is particularly suitable for the mesh implementation. To implement each step of the multi-step iteration requires

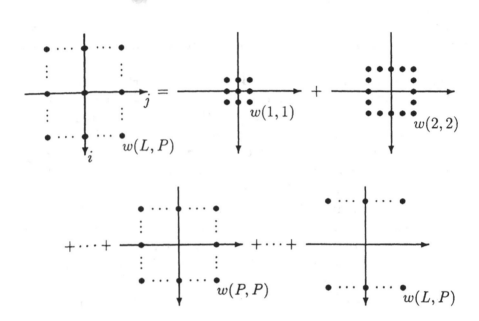

Figure 6: Decomposition of the imnpulse response of the restoration filter

considerably less time than the single-step iteration. More specifically, this reduction is proportional to the largest dimension of the support of the impulse response of $w(i,j)$ in Eq. (2.9).

4.1 Derivation

In this section we assume that the support of the impulse response of the composite filter $w(i,j)$ in Eq. (2.9) is rectangular of size $(2L+1) \times (2P+1)$ pixels where $L \geq P$. This does not represent a deviation from the presentation in Sec. 3 where $w(i,j)$ had a square support, since any square region of support can be treated as rectangular by padding it with the appropriate number of zeros.

We propose the following additive decomposition of $w(i,j)$

$$w(i,j) = w_1(i,j) + w_2(i,j) + \ldots + w_L(i,j), \qquad (4.1)$$

where the functions $w_l(i,j), l = 1, \ldots, L$ are depicted in Fig. 6. Then iteration (2.9) takes the form

$$
\begin{aligned}
x_0(i,j) &= \beta a(-i, -j) \star \star g(i,j) \\
x_k(i,j) &= w_1(i,j) \star \star x_{k-1}(i,j) \\
&+ w_2(i,j) \star \star x_{k-2}(i,j) + \ldots \\
&+ w_2(i,j) \star \star x_{k-L}(i,j) + f(i,j)
\end{aligned}
\tag{4.2}
$$

or the following matrix-vector form

$$
\begin{aligned}
x_0 &= \beta A^T g \\
x_k &= W_1 x_{k-1} + W_2 x_{k-2} + \ldots \\
&+ W_P x_{k-P} + \ldots + W_L x_{k-L} + f,
\end{aligned}
\tag{4.3}
$$

where the sequences $w_1(i,j), \ldots, w_L(i,j)$ are used in forming the matrices $W_1 \ldots, W_L$, respectively. Each of these matrices is block banded, where each block is a banded matrix.

4.2 Convergence

The convergence of the single-step algorithm does not in general guarantee the convergence of the multi-step algorithm. In the following, sufficient conditions for the convergence of the multi-step algorithm are discussed. A sufficient condition for convergence of iteration (4.3) to a unique solution for any $x_0, \ldots, x_{-L+1}$ is that [16, 31]

$$
\sum_{l=1}^{L} \|W_l\| < 1.
\tag{4.4}
$$

Since the contraction condition (4.4) is norm dependent [31], all useful norms should be used in verifying (4.4). The form of this condition with the use of the l_1, l_2 and l_∞ matrix norms is considered next.

Matrices $W_l, l = 1, \ldots, L$ are block Toeplitz, therefore they are asymptotically equivalent to block circulant matrices [1]. The eigenvalues of the $W_l's$ (real eigenvalues since the $W_l's$ are symmetric) are the values of the 2-D Discrete Fourier Transform (DFT) of the $w_l(i,j), l = 1, \ldots, L$ in equation (4.1). In other words, in trying to verify (4.4) with the use of l_2 norms, we need to find the maximum of each individual 2-D DFT. That is, condition (4.4) is translated into

$$
\sum_{l=1}^{L} max_{n_1, n_2} W_l(n_1, n_2) = \sum_{l=1}^{L} \rho(W_l) < 1
\tag{4.5}
$$

where $\rho(W_l)$ denotes the spectral radius of W_l, and $W_l'(n_1, n_2)$ is the 2-D DFT of $w_l(i,j)$ with $n_1 = 2\pi/N_1$ and $n_2 = 2\pi/N_2$, where a $N_1 \times N_2$ size DFT is assumed, with $N_1 \geq (2L+1)$ and $N_2 \geq (2P+1)$. Verification of equation (4.5) is not an analytically straightforward task, since each of the $W_l(n_1, n_2)$ obtains its maximum value at a different (n_1, n_2) point. Therefore, no general conclusion can be reached and condition (4.5) needs to be numerically verified for a given $w(i,j)$.

A sufficient condition which is simpler to verify can be obtained by considering the l_1 and l_∞ norms. In this case

$$||W||_1 = ||W||_\infty = \sum_{l=1}^{L} ||W_l||_1 = \sum_{l=1}^{L} ||W_l||_\infty, \qquad (4.6)$$

due to the symmetry of the W_l's and the fact that $W = \sum_l W_l$. Condition (4.4), due to equation (4.6) results in the following expression

$$\begin{aligned} ||W||_1 &= ||W||_\infty \\ &= \sum_j \sum_j |w(i,j)| < 1, \end{aligned} \qquad (4.7)$$

which is very straightforward to be checked for a given $d(i,j), c(i,j), \beta$ and α. In verifying conditions (4.5) and (4.7) it should be kept in mind that β is the only free parameter.

If W is a contraction with respect to the l_∞ norm, that is, $||W||_\infty < \delta$, where $\delta < 1$, a more powerful result than the convergence of iteration (4.3) holds [32]. More specifically an asynchronous distributed implementation of the iterative algorithm (2.6), as described in Ref. [32] is proved to converge to a unique solution. Furthermore, the notion of a P-contraction [31, 32] can also be used in establishing sufficient conditions for the convergence of the general case of a distributed asynchronous iterative algorithm. However, if a mapping is a P-contraction, then it is a contraction mapping with respect to the l_∞ norm [31] and the previously mentioned result holds.

Provided that an iterative algorithm converges, the next important question is how fast is this convergence. For the single-step iteration of equation (2.6), the convergence is linear, as expressed by Eqs. (2.7) and (2.8). If A is invertible, then c in Eq. (2.8) is equal to the spectral radius of W, denoted by $\rho(W)$. Then $||x_k - x^*|| \leq \rho(W)^{k+1}$, where x^* is the fixed point of iteration (2.6). The multi-step iterative algorithm has also linear convergence. It is shown ([31] p. 354) that if condition

(4.4) is satisfied then the asymptotic convergence factor is strictly less than one and it is equal to the spectral radius of the following matrix

$$
H = \begin{bmatrix}
W_1 & W_2 & \cdots & \cdots & W_L \\
I & 0 & \cdots & \cdots & 0 \\
& \ddots & \ddots & & \vdots \\
& & & & \vdots \\
& & \ddots & \ddots & \vdots \\
0 & & & I & 0
\end{bmatrix}
\tag{4.8}
$$

Depending on the specific $d(i,j), c(i,j), \beta$ and α, the spectral radii of W and H can be computed in order to compare the rate of convergence of the multi-step and the single-step iteration.

4.3 Mesh Implementation

In contrast to the implementation described in Sec. 3.3, both communication and computation take place in every clock cycle in the implementation of the multi-step iteration. Each clock cycle k, which is the time interval $[k, k+1)$, is assigned to be divided into two phases. Computation (described below) takes place in the first phase which is the interval $[k, k+\Delta)$, with $0 < \Delta < 1$. In the second phase, interval $[k+\Delta, k+1)$, the communication (to be specified below) takes place. Let $M_x^{(i,j)}(l, m; t)$ denote the contents of the $(l, m)th$ location of the memory matrix $M_x^{(i,j)}$ at any time t. The computation performed in cycle k by processor (i, j) is

$$
\sum_{m=-P}^{+P} \sum_{l=-P}^{+P} M_x^{(i,j)}(l, m; k) M_w^{(i,j)}(l, m) + f(i, j),
\tag{4.9}
$$

where for convenience we have assumed that $L = P$. The result of the above computation is copied onto the central location of $M_x^{(i,j)}$, by the end of the computation phase (i.e., time $k + \Delta$). This value remains unchanged for the rest of the cycle k. All the other locations (there are $(2P + 1)^2 - 1 = 4P^2 + 4P$ of them) get altered during the communication phase. The values transmitted by the neighbors during the communication phase of cycle k are copied onto these locations. For example,

$$
M_x^{(i+1,j)}(0, 0; k + \Delta) \rightarrow M_x^{(i,j)}(1, 0; k + 1).
\tag{4.10}
$$

If there is no duplication of messages in the communication sequence, then each of the eight neighbors transmits $(4P^2 + 4P)/8$ messages (the contents of a location) to a processor. In the following we describe a communication pattern that avoids duplication.

For each processor (i, j), divide the matrix $M_x^{(i,j)}$ into eight octants, as shown in Fig. 7a. Thus for processor (i, j), octant l is the set of memory locations $M_x^{(i,j)}(l, m)$ where $-P \leq l < 0$ and $l > m$; for octant 2, the index values are $-P \leq l < 0$ with $l \leq m$ and so on.

In the communication phase of each cycle, each processor transmits the values (found at the end of the computation phase) in an octant to a specific neighbor. Thus, each octant goes to one of its neighbors, according to the following rule. Octant 1 is transmitted to the D neighbor, which we shall denote by $1 \rightarrow D$. The other transmission rules are: $2 \rightarrow DL; 3 \rightarrow L; 4 \rightarrow UL; 5 \rightarrow U; 6 \rightarrow UR; 7 \rightarrow R$; and $8 \rightarrow DR$. In other words, the octants 1 through 8 are transmitted clockwise, one per neighbor, to each of the neighbors starting from $1 \rightarrow D$. The messages received by processor (i, j) are shown in Fig. 7b. Note that the outermost locations of an octant need not be transmitted. In every cycle, the received messages are immediately copied onto the relevant local memory locations. For example, when the U neighbor sends the contents of its location (l, m), it is copied onto location $(l - 1, m)$ of the receiving processor, i.e.

$$M_x^{(i-1,j)}(l, m; k + \Delta) \rightarrow M^{(i,j)}(l - 1, m; k + 1). \qquad (4.11)$$

The above communication pattern includes vertical, horizontal, and diagonal communication. To see the information transfer implied in this communication pattern, consider the undirected graph of a two-dimensional array of nodes, obtained from associating a node with each processor, and an edge connecting two neighboring processors which exchange messages. Then, as a consequence of the above described communication, at the end of any cycle k,

$$M_x^{(i,j)}(0, 0; k) = x_k(i, j) \qquad (4.12)$$

and

$$M_x^{(i,j)}(l, m; k) = x_{k-e}(i + l, m + j) \qquad (4.13)$$

where $-P \leq l \leq P, -P \leq m \leq P$, and where e is the shortest distance in the graph between nodes (i, j) and $(i + l, m + j)$, which equals the

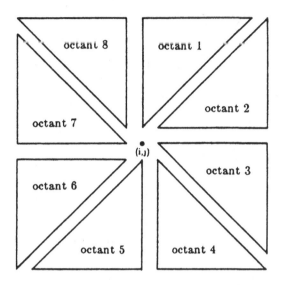

a. Octant assignment of local memory

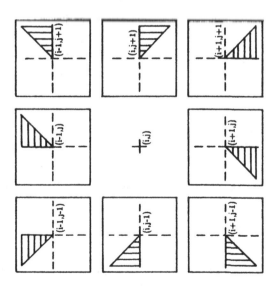

b. Message transmission/reception pattern

Figure 7: Matrix $M_x^{(i,j)}$

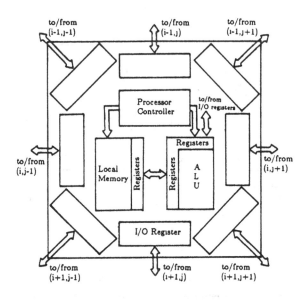

Figure 8: A block diagram of a processor (i, j)

maximum of l and m. In other words, the restored value from a processor which is at a distance e away, arrives after a delay of e cycles. This implies that the evolution of the restoration process can be represented by the multi-step iteration (4.3) analyzed in subsection 4.1. A block diagram of the processor in the architecture is given in Fig. 8. The various blocks are self-explanatory. The multi-step iteration can also be implemented with 4-neighbor communication (transmitting quadrants instead of octants), using the ideas of the 8-neighbor communication described above. In this case the communication pattern is determined by the partitioning of $M_x^{(i,j)}$ shown in Fig. 2. Note that when diagonal are used in the mesh the computation is reduced at most by a factor of 2. Thus, in our asymptotic analysis the time complexity is not affected.

4.4 Performance

In the architecture explained above, each processor transmits a total of $4P^2 + 4P = (Q - 1)$ messages per cycle, where $Q = (2P + 1)^2$ and the length of each message equals the number of bits used to represent a pixel. (If the template is rectangular i.e, $P \neq L$, then the number of messages per cycle is $4PL + 2(P + L)$.) The local memory requirement is also proportional to the size of the template, being equal to $2Q$.

In the above implementation, the speedup of the restoration process, in terms of the gain in rate of convergence, is proportional to the number of processor (N_T). Notice that if only a single processor were to be employed, then the computation involved in each step of iteration (2.6) would require N_T clock cycles (assuming that the convolution, such as equation (4.9) required for each pixel can be done in one clock cycle). Thus, significant speedups are possible by employing a large number of processors. In general, if γ_2 represents the spectral radius of H in equation (4.8), then the rate of convergence of the multi-step iteration (implementation described above) with N_T parallel processors is $\gamma_2^{N_T}$. In other words, if N_T clock cycles constitute one time period, then the norm of the error at the end of k time periods $||x_k - x^*||$ is $O(\gamma_2^{kN_T})$. In contrast, the rate of convergence of the single-step implementation with an array of N_T processors is given by $\gamma_1^{(N_T/L)}$, where L is the length of the (square) template and γ_1 is the spectral radius of W in equation (2.6). In this case $||x_k - x^*|| = O(\gamma_1^{k(N_T/L)})$. Therefore, in situations where the template size is large, one can expect the multi-step iteration implementation to be considerably faster than the single-step iteration implementation (even if γ_2 is smaller than γ_1).

There are N^2 processor each with constant area. Thus area is $O(N^2)$. It takes one unit of time to multiply two numbers in the word model. Thus, $AT^2 = O(N^2)$ in the word model. As discussed in Sec. 3.3, we obtain the following result in the bit model (for both diagonal and non-diagonal communication).

Theorem 4: The multistep algorithm works in T_1^2 time has area $O(\frac{N^2}{T_1^2})$, for $T_1 \epsilon [O(logb), O(\sqrt{b})]$, where b is the operand size.

5   DISCUSSION AND CONCLUSIONS

Mesh, pyramid, and mesh of pyramids implementations for an iterative image restoration algorithm have been proposed. These implementations are based on a single step iterative algorithm. The efficiency of the VLSI algorithms is judged by establishing lower bounds on functions which capture an area-time tradeoff. The lower bounds for AT^2 which have been obtained for these three architectures, explicitly indicate the dependence on Q, the size of the filter support, and b, the length of the operands. Clearly, the time of the MOP implementation is by far superior than the mesh implementation. However, mesh has an attractive VLSI implementation due to regularity.

As described in the text, our AT^2 bounds are away from the lower bound by a factor of $O(P^2)$. The derivation of architectures with optimal AT^2 is currently under investigation. It is also conceivable that by using problem transformation techniques described in [22], tighter lower bounds can be obtained. The VLSI implementation of iterative restoration algorithms with higher convergence rates than the ones presented here [13], and the investigation of convergence and implementation of asynchronous iterative algorithms, are topics for future research.

References

[1] H. C. Andrews and B. R. Hunt, *Digital Image Restoration*, Prentice-Hall, 1977.

[2] R. W. Schafer, R. M. Mersereau and M. A. Richards, "Constrained Iterative Restoration Algorithms", *Proc. IEEE,* Vol. 60, pp., 432-450, April 1981.

[3] A. K. Katsaggelos, "Constrained Iterative Image Restoration Algorithms", *Optical Engineering*, special issue on Visual Communications and Image Processing, Vol. 28, No. 7, pp. 735-748, July 1989.

[4] A. K. Katsaggelos, J. Biemond, R. W. Schafer and R. M. Mersereau, "A Regularized Iterative Image Restoration Algorithm", *IEEE Trans. Acoust., Speech, Signal Processing*, Vol. 39, No. 4, April 1991.

[5] J. L. Potter, "The STARAN Architecture and its Application to Image Processing and Pattern Recognition Algorithms", *Proc. Nat. Comp. Conf.,* 1978.

[6] K.E. Batcher, "Design of a Massively Parallel Processor", *IEEE Trans. Comput.,* Vol. 29, pp. 836-840, 1980.

[7] W. D. Hillis, "The Connection Machine: A Computer Architecture Based on Cellular Automata", *Physica*, Vol. 10D, pp. 213-228, 1984.

[8] S. Y. Kung, H. J. Whitehouse and T. Kailath, editors, *VLSI and Modern Signal Processing,* Prentice-Hill, 1985.

[9] R. J. Offen, editor, *VLSI Image Processing*, McGraw-Hill, 1985.

[10] H. T. Kung and S. W. Song, "A Systolic 2-D Convolution Chip", *IEEE Com. Soc. Workshop on Computer Architecture for Pattern Analysis and Image Database Management*, pp. 159-160, Nov. 1981.

[11] H. T. Kung "Why Systolic Architectures?", *IEEE Comput.,* Vol. 15, No. 1, pp. 37-46, Jan. 1982.

[12] I.-C. Wu *Area-Time Tradeoffs in VLSI Algorithms*, M.Sc. Thesis, National Taiwan University, 1984.

[13] A. K. Katsaggelos and S. N. Efstratiadis, "A Class of Iterative Signal Restoration Algorithms", *IEEE Trans. Acoust., Speech, Signal Processing,* Vol. 38, No. 5, pp. 778-786, May 1990.

[14] D. C. Youla and H. Webb, "Image Reconstruction by the Method of Convex Projections, Part 1-Theory", *IEEE Trans. on Medical Imaging,* Vol. MI-1, No. 2, pp. 81-94, Oct. 1982.

[15] A. K. Katsaggelos, S. P. R. Kumar and M. R. Samatham, "VLSI Implementation of an Iterative Image Restoration Algorithm", *Proc. 1986 Int. Conf. Sys. Man. and Cybern.,* Atlanta, GA, pp. 313-318, Oct. 1986.

[16] A. K. Katsaggelos and S. P. R. Kumar, "Single and Multistep Iterative Image Restoration and VLSI Implementation", *Signal Processing, Vol. 16, No. 1, pp. 29-40, Jan. 1989.*

[17] A. K. Katsaggelos, S. P. R. Kumar and M. Sarrafzadeh, "Parallel Processing Architectures for Iterative Image Restoration", *Proc. of 1989 Int. Conf. on Acoust., Speech, and Signal Processing,* pp. 2544-2547, Glasgow, Scotland, May 1989.

[18] C. D. Thompson, *A Complexity Theory for VLSI,* Ph.D. Thesis, Department of Computer Science, Carnegie-Mellon University, Pittsburgh, PA, 1980.

[19] H. Abelson and P. Andreae, "Information Transfer and Area-Time Trade-offs for VLSI Multiplication", *Communications of the ACM,* 23, pp. 20-22, 1980.

[20] R. P. Brent and H. T. Kung, "The Area-Time Complexity of Binary Multiplication", *Journal of the ACM,* 28, pp. 521-534, 1981.

[21] G. Bilardi, M. Pracchi, and F. Preparata, "A Critique and Appraisal of VLSI Model of Computation", *Proc. CMU Conference on VLSI Systems and Computations,* 1981.

[22] S. W. Hornick and M. Sarrafzadeh "On Problem Transformability in VLSI", *Algorithmica,* 2, pp. 97-111, 1987.

[23] C. D. Thompson,"Area-Time Complexity for VLSI", *Proceedings of the 11th Annual ACM Symposium on the Theory of Computing,* Atlanta, GA, pp. 81-88, 1979.

[24] G. Bilardi and F. P. Preparata, "Tessellation Techniques for Area-Time Lower Bounds with Application to Sorting", *Algorithmica,* Vol. 1, No. 1, pp. 65-91, 1986.

[25] D. E. Dudgeon and R. M. Mersereau, *Multidimensional Digital Signal Processing,* Prentice-Hall, 1984.

[26] K. Mehlhorn and F. Preparata, "Area-Time Optimal VLSI Integer Multiplier with Minimum Computation Time", *Information and Control,* Vol. 58, pp. 137-156, 1983.

[27] G. Bilardi and M. Sarrafzadeh, "Optimal VLSI Circuits for Discrete Fourier Transform", in *Advances in Computing Research,* Vol. 4, (F. P. Preparata, Editor), JAI Press, Greenwich, CT., pp. 87-101.

[28] C. R. Dyer, "A VLSI Pyramid Machine for Hierarchical Parallel Image Processing", *Proc. of Pattern Recognition and Image Processing Conference,* TX, pp. 381-386, 1981.

[29] J. K. Chang, O. H. Ibarra, T. C. Pong, and S. M. Sohn, "Two-Dimensional Convolution on a Pyramid Computer", *Proc. Int. Conf. on Parallel Processing,* pp. 780-782, 1987.

[30] M. Sarrafzadeh, S.P.R. Kumar, and A. K. Katsaggelos, "Parallel Architectures for an Iterative Image Restoration Algorithm", *Proc. Int. Symp. on Circuits and Systems,* pp. 2605-2609, New Orleans, LA, May 1990.

[31] J. M. Ortega and W. C. Rheinboldt, *Iterative Solutions of Non-linear Equations in Several Variables,* Academic Press, NY., 1970.

[32] D. P. Bertsekas, "Distributed Asynchronous Computations of Fixed Points", *Math. Programming,* Vol. 27, pp. 107-120, 1983.

2

Perfect Shuffle Communications
In Optically Interconnected Processor Arrays

Zicheng Guo and Rami G. Melhem

Departments of Electrical Engineering and Computer Science
University of Pittsburgh, Pittsburgh, PA 15261

Abstract

Two approaches are presented to perform the perfect shuffle communication in optically interconnected processor arrays. The arrays are interconnected via row and column optical busses, with optical switches placed at the intersections of row and column busses. The two approaches differ in the time they take and the switching complexity they require. They allow all algorithms, including those for digital signal processing (e.g., FFT), which utilize the shuffle-exchange communication structure to be efficiently executed on the optically interconnected processor arrays.

1. Introduction

Because of its efficiency, the perfect shuffle communication has been used in the design of many parallel algorithms, including FFT, sorting, matrix operations, and image computation [4, 12, 14, 18, 21]. It is desirable to be able to emulate the perfect shuffle communication structure on a given parallel computer since such emulation allows all algorithms designed for the perfect shuffle to be executed on the parallel computer. Optical implementations of the perfect shuffle using free space techniques have been proposed by several authors [2, 11]. In this paper we show how the perfect shuffle communication can be accomplished using guided optics in optically interconnected processor arrays. Due to their capability of pipelining messages on optical busses, these hybrid optical-electronic multiprocessor architectures, called *Array Processors with Pipelined Busses* (APPB), have been shown to achieve an asymptotically linear (in number of processors on the bus) improvement in communication bandwidth over conventional multiprocessor architectures with nearest neighbor or exclusive access bus interconnections [7, 8].

Message pipelining on optical busses is possible because optical signals have two unique properties which are not shared by their electronic counterpart.

This work was partially supported by Air Force grant AFOSR-89-0469 and NSF grant MIP-8901053.

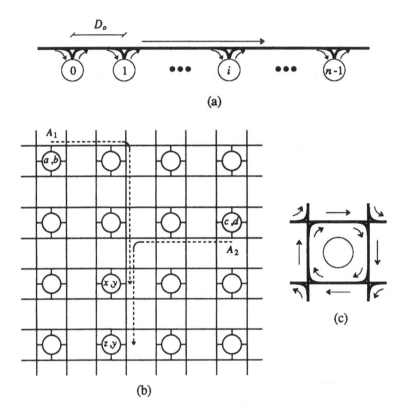

Figure 1. (a) A processor array connected with a single optical bus.
(b) APPB with switches where each processor is coupled to four
optical busses. (c) Switch connections at each intersection of
row and column busses.

Namely, optical signal's propagation is unidirectional and has a precisely predi-
catable path delay per unit distance. Figure 1(a) illustrates an array of n electronic processors connected by an optical bus (waveguide), where each processor is coupled to the bus with two passive directional optical couplers [9, 20],
one for injecting optical signals on the bus, and the other for receiving from the
bus. As in the case of electronic busses, each node j communicates with any
other node i by sending a message to i through the common bus. However,
because optical signals propagate unidirectionally, a node j in the system of
Figure 1(a) may send a message to another node i only if $i > j$. A message on
an optical bus consists of a sequence of optical pulses, each having a width w in
seconds. The existence of an optical pulse of length w represents a binary bit 1,
and the absence of such a pulse represents a 0. For convenience, let D_o be the
optical distance between each pair of adjacent nodes and τ be the time taken for
an optical signal to traverse the optical distance D_o. To transfer a message from

a node j to node i, $i > j$, the sending node j writes its message on the bus. After a time $(i-j)\tau$, the message will arrive at the receiving node i, which then reads the message from the bus. To facilitate our discussion, for the system in Figure 1(a) we define $n\tau$ as a *bus cycle*, and correspondingly τ as a *petit cycle*.

Assume the system of Figure 1(a) is synchronized such that every processor writes its message on the bus at the beginning of a bus cycle and that the optical distance D_o is larger than the message length bwc_g, where b is the number of binary bits in each message and c_g is the speed of light in the waveguide. Then all the processors can send their messages on the bus *simultaneously*, and all the messages will then travel from left to right on the bus in a pipelined fashion without collision. Here by collision we mean that two messages sent by two distinct processors arrive at some point on the bus simultaneously. This is in contrast to an electronic bus, where writing access to the bus is *exclusive*. In cases where the communication pattern is known to the receiver, that is, the receiver knows who the sender is, which is true for most communications/computations in SIMD architectures [19], a *wait* register in each processor may be programmed such that it indicates the time at which the processor should read its message from the bus.

Several other addressing mechanisms can be used for transferring messages on the optical bus. For example, a *skip* register may be used to count the number of messages to be skipped before reading the right message [13]. This mechanism relaxes the requirements for timing accuracy and for equal distance between each pair of consecutive processors on the bus. In cases where the communication pattern is unknown to the receiver, the destination address can be included in each message. Coincident pulse techniques [3, 10] provide a mechanism for an all-optical encoding/decoding of destination addresses.

Connecting all processors in a system with a single optical bus, as shown in Figure 1(a), has the disadvantage that a message transfer incurs $O(N)$ time delay in an N-processor system. This delay is reduced to $O(\sqrt{N})$ in the two-dimensional APPB [7], where each processor is connected to four optical busses as discussed in the next section.

2. Array Processors with Pipelined Busses Using Switches

In the two-dimensional APPB architecture each processor is coupled to four optical busses, with two horizontal busses for passing messages horizontally in opposite directions, respectively, and two vertical busses for passing messages vertically in a similar way [7, 8]. The two-dimensional APPB architecture achieves a significant reduction in the length of a bus cycle, however, it may take two steps for two processors to communicate with each other. To be more specific, a message has to be sent to and buffered at an intermediate processor in the first step, and then relayed from that intermediate processor to its destination in the second step. Such message relay reduces the communication efficiency since it requires an optical-electronic-optical information conversion at the intermediate processor.

One way of dealing with this disadvantage of message relay is to use 2×2 optical switches, e.g., Ti:LiNbO$_3$ switches [1, 17], to connect row and column busses. A 2×2 optical switch has two inputs, I_1 and I_2, which, depending on the value of the control signal C, can be directly or cross connected to two outputs O_1 and O_2, respectively (see Figure 2). These switches have been used to implement interconnection networks [5, 23], memories [16], and counters [22]. In this paper they are used to switch an optical signal, say, from a row bus directly to a column bus without requiring a relay by an intermediate processor. The architecture of APPB with switches is schematically drawn in Figure 1(b), where switch connections at each processor are shown in (c). For an $m \times n$ APPB with switches a bus cycle is defined as $(m + n)\tau$.

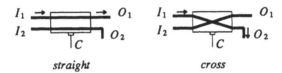

straight   *cross*

Figure 2. A 2×2 optical switch and its state definition.

In APPB with switches, a switch may assume one of the two states *straight* and *cross* as defined in Figure 2. Initially all switches are in state *straight*. When a message switching is desired at some processor, a switch at that processor must be set to the *cross* state. The state of a switch at a processor (x, y) in an $m \times n$ APPB with switches is determined by a variable $S_{ij}(x, y)$, $0 \le x < m$, $0 \le y < n$, and $i, j \in \{R, L, D, U\}$, where R, L, D, and U stand for rightward, leftward, downward, and upward, respectively. For example, $S_{RD}(x, y)$ is used to specify the control of the switch which guides optical signals in rightward-to-downward direction at processor (x, y). The value of $S_{ij}(x, y)$ is a tuple (λ, μ), where the integer λ specifies the time, in number of petit cycles and relative to the beginning of a bus cycle, at which the switch is set to *cross*, and the integer μ determines the time period, again in number of petit cycles, during which the switch should remain *cross*. We assume that two switches are used at each intersection of row and column busses. For example, at the intersection of the rightward row bus and the downward column bus, the two switches are $S_{RD}(x, y)$ and $S_{DR}(x, y)$. In this case one input of each switch is left unused. Figure 2 shows the implementation of $S_{RD}(x, y)$ where input I_2 is not used. Note that if $S_{RD}(x, y) = S_{DR}(x, y)$, then the two switches at the intersection of the rightward row bus and the downward column bus may be combined into a single switch, thus reducing the hardware requirement by half.

A switch may be controlled in three different modes:

1)   The switch is set to *cross* at the beginning of a bus cycle and remains at *cross* throughout the entire bus cycle.

2) The switch is set to *cross* in the middle of a bus cycle and remains at *cross* thereafter in that bus cycle.

3) The switch is set to *cross* in the middle of a bus cycle, remains at *cross* for a few petit cycles, and is then set back to its initial state before the end of that bus cycle.

From the definition, for modes 1) and 2), we have $\lambda + \mu = m + n$, and for mode 3), $\lambda + \mu < m + n$. These switching modes have different implementation complexity and routing flexibility. Considering the number of switchings in the *middle* of a bus cycle, the three modes require zero, one, and two switchings, respectively. Thus the switching complexity of the three modes increases from 1) to 3). With the increasing complexity, the flexibility of these modes, in terms of routing messages, also increases.

The switching mode 1) can be used to accomplish some simple and useful communication tasks, for example, matrix transpose [6]. In this paper, we present two approaches to perform the perfect shuffle communication on APPB with switches using modes 2) and 3). The first approach requires mode 2) and takes two bus cycles. The second uses one bus cycle but requires the more complex mode 3).

Message switching between row and column busses may cause message collisions, and extra care must be taken to ensure collision-free message routing when designing communications for APPB with switches. A sufficient and necessary condition for collision free communication in APPB with switches is given in the following Lemma [7].

Lemma. Assume that the optical distance D_o between two consecutive processors is larger than the message length and that all processors start sending their messages simultaneously. Then, two messages sent by two distinct processors (a, b) and (c, d), respectively, passing processor (x, y) on the same bus will collide if and only if

$$|a - x| + |b - y| = |c - x| + |d - y| \tag{1}$$

As an example of message collision in APPB with switches, the two messages A_1 and A_2 in Figure 1(b) traveling from (a, b) and (c, d) to (x, y) and (z, y), respectively, are colliding on the downward bus at processor (x, y). Note that for the Lemma to hold, it is necessary that the two messages pass processor (x, y) on the same bus. For example, if message A_1 in Figure 1(b) is switched at processor (c, y) from the downward bus to a row bus, it will not pass processor (x, y). As a result, A_1 will not collide with A_2 even if Eq. (1) holds true. In the following we present approaches to performing the perfect shuffle communication in APPB with switches and show that they are collision-free.

3. A Two-Cycle Approach to the Perfect Shuffle Communication

Consider an APPB with switches of size $n \times n$ where the processors are numbered using row major indexing. To perform the perfect shuffle communication in this APPB, a processor i will communicate with processor $Shuffle(i)$, where $Shuffle(i) = 2i$ if $0 \le i < n^2/2$, and $Shuffle(i) = 2i \bmod n^2 + 1$ otherwise [18]. In terms of row/column positions, a processor (x, y) will communicate with $Shuffle[(x, y)]$ defined as follows [15].

$$Shuffle[(x,y)] = \begin{cases} (2x, 2y), & 0 \le x < \frac{n}{2}, 0 \le y < \frac{n}{2} \\ (2x+1, 2y \bmod n), & 0 \le x < \frac{n}{2}, \frac{n}{2} \le y < n \\ (2x \bmod n, 2y+1), & \frac{n}{2} \le x < n, 0 \le y < \frac{n}{2} \\ (2x \bmod n+1, 2y \bmod n+1), & \frac{n}{2} \le x < n, \frac{n}{2} \le y < n \end{cases} \quad (2)$$

From this definition, we can divide the messages in the shuffle communication into four sets, M_i, $1 \le i \le 4$, according to their source position in the four quadrants specified in the above definition. That is, M_i is the set of messages with source processors from quadrant i.

For the perfect shuffle communication, we choose to transmit messages such that they will propagate in counterclockwise direction: Messages from M_1, M_2, M_3, and M_4 are transmitted in downward-to-rightward, leftward-to-downward, rightward-to-upward, and upward-to-leftward direction, respectively. As an example, a message in M_1 will be written on the downward bus in its source column, and then switched rightward at its destination row. Typical message flows are shown in Figure 3(a) for messages in M_1. Note that messages from M_1 have their destinations scattered in all the four quadrants. For simplicity, we will depict the message flows as in Figure 3(b). Similar figures can be drawn for other message sets. It should be clear that the messages in M_1, M_2, M_3, and M_4 will be switched by switches S_{DR}, S_{LD}, S_{RU}, and S_{UL}, respectively.

To show how each switch $S_{ij}(x, y)$ should be controlled, we consider $S_{DR}(x, y)$. From definition (2), a message in M_1 from source processor (X, y) has its destination row at $2X$. Thus the distance covered on the downward column bus by the message is $2X - X = X$, which is numerically equal to the time, in number of petit cycles, at which $S_{DR}(2X, y)$ should be set to $cross$. That is, $S_{DR}(2X, y) = (X, \mu)$. Or equivalently, $S_{DR}(x, y) = (x/2, \mu)$, where x is even. Using switching mode 2), we have $\mu = 2n - x/2$. Similarly other switch controls can also be determined. These are given in the following.

$$S_{DR}(x,y) = (\frac{x}{2}, 2n - \frac{x}{2}), \quad x \text{ even}, 0 \le y < \frac{n}{2} \text{ (for } M_1) \quad (3a)$$

$$S_{LD}(x,y) = (\frac{n-y}{2}, 2n - \frac{n-y}{2}), \quad 0 \le x < \frac{n}{2}, y \text{ even (for } M_2) \quad (3b)$$

$$S_{RU}(x,y) = (\frac{y+1}{2}, 2n - \frac{y+1}{2}), \quad \frac{n}{2} \le x < n, y \text{ odd (for } M_3) \quad (3c)$$

$$S_{UL}(x,y) = (\frac{n-x-1}{2}, 2n - \frac{n-x-1}{2}), \quad x \text{ odd}, \frac{n}{2} \le y < n \text{ (for } M_4) \quad (3d)$$

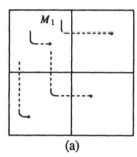 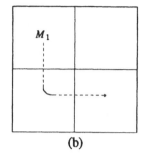

(a)   (b)

Figure 3. Typical message flows for M_1 in the perfect shuffle communication. Arrowed curves start from the source quadrants of messages and end in the destination quadrants. (a) Messages in M_1 have their destinations in all the four quadrants. (b) A simplified representation of (a).

Thus these switch controls are defined such that if only a single processor (x, y) is transmitting a message to processor $Shuffle [(x, y)]$, then the message will be correctly sent to a bus at $Shuffle [(x, y)]$. To receive the message, the value to be stored in the *wait* register at processor $Shuffle [(x, y)]$ can be computed based on the Manhattan distance between (x, y) and $Shuffle [(x, y)]$, which can be determined from definition (2).

The perfect shuffle communication can be performed in two bus cycles. In the first bus cycle, processors in quadrants 1 and 4 send their messages in counterclockwise direction. The switch settings are as defined in Eqs. (3a) and (3d). In the second bus cycle, processors in quadrants 2 and 3 send their messages in counterclockwise direction. The switch settings are as defined in (3b) and (3c). $S_{ij}(x, y) = (0, 0)$ if not specified, that is, the switch will stay in *straight* throughout the entire bus cycle. Typical message flows in the perfect shuffle communication using this approach is shown in Figure 4.

Proposition 1. The perfect shuffle communication using two cycles is collision-free.

Proof. Our proof is given for the first bus cycle. The case for the second cycle follows similarly. It can be checked that messages from M_1 and M_4 do not travel in the same row or column. Thus these two sets of messages cannot collide. Now consider the messages from M_1 which are switched to the same row and thus might collide with one another after being switched. Let B_1 and B_2 be two such messages which are propagating from two distinct processors (a, b) and (e, f), respectively, to the same destination row x, where x is even. Then from Eq. (2) these messages are from the same source row $x/2$. Thus, we have $a = e = x/2$ and $b \neq f$. Assume that both B_1 and B_2 will pass processor (x, y) on the rightward bus. Then we have $b < y$ and $f < y$. Given these values, Eq. (1) does not

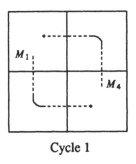
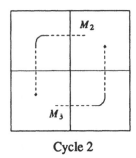

Cycle 1   Cycle 2

Figure 4. Typical message flows in the perfect shuffle communication
using two cycles. Arrowed curves start from the source quadrants of
messages and end in the destination quadrants.

hold. That is, B_1 and B_2 do not collide. This completes the proof of Proposition 1. $\square$

This Proposition tells us that in two bus cycles, the perfect shuffle communication can be performed in APPB with switches using switching mode 2). The interested reader may verify that if mode 1) had been used in this approach, message collisions would have occurred. Thus the simpler switching mode 1) is not sufficient for the powerful shuffle communication. Does the more complex switching mode 3) have any advantage over mode 2) in terms of performing the shuffle communication? The answer is positive. In the next section we show that using switching mode 3), the task can be accomplished in one bus cycle.

4. A One-Cycle Approach to the Perfect Shuffle Communication

In this section we show how the perfect shuffle communication can be performed on APPB with switches using a single bus cycle. The idea is to have all processors transmit their messages in the same bus cycle, instead of in two cycles, but use the more complex switching mode 3) so that message collisions cannot occur.

At the beginning of a bus cycle, all processors transmit their messages in counterclockwise direction. The switch controls are as follows.

$$S_{DR}(x,y) = (\frac{x}{2}, 1), \quad x \text{ even}, 0 \leq y < \frac{n}{2} \text{ (for } M_1) \tag{4a}$$

$$S_{LD}(x,y) = (\frac{n-y}{2}, 1), \quad 0 \leq x < \frac{n}{2}, y \text{ even (for } M_2) \tag{4b}$$

$$S_{RU}(x,y) = (\frac{y+1}{2}, 1), \quad \frac{n}{2} \leq x < n, y \text{ odd (for } M_3) \tag{4c}$$

$$S_{UL}(x,y) = (\frac{n-x-1}{2}, 1), \quad x \text{ odd}, \frac{n}{2} \leq y < n \text{ (for } M_4) \tag{4d}$$

If the setting for a switch $S_{ij}(x, y)$ is not specified, then $S_{ij}(x, y) = (0, 0)$.

Note that in this approach each switch will remain at the *cross* state for one petit cycle, that is, $\mu = 1$. The reason for this will become clear in the proof of Proposition 2 presented later. Typical message flows in the perfect shuffle communication using one cycle is shown in Figure 5.

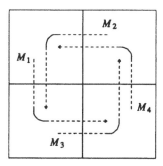

Figure 5. Typical message flows in the perfect shuffle communication using one cycle. Arrowed curves start from the source quadrants and end in the destination quadrants.

To prove that this approach guarantees collision-free communication, we need only to consider messages from two adjacent quadrants. Let us look in more detail at how messages in M_1 and M_2 may possibly interfere with one another (messages from other sets can be looked at similarly). From definition (2), messages in M_2 have their destination processors at only even columns. These messages will first propagate to the left on their source row busses, and then be switched downward at even columns. While messages in M_1 will first propagate downward in their source columns and then be switched rightward at their destination rows (even). Since messages in M_1 are from the first quadrant, that is, their source columns are from the first quadrant, messages from M_1 and M_2 can possibly interfere with each other on the downward bus only at columns y, where $0 \le y < n/2$ and y is even. (See Figure 6.) Observe that messages in M_1 are injected directly on the downward bus at column y, while messages in M_2 are injected on their respective row busses at certain distances from column y and it will take certain amount of time for them to reach column y. Thus for a message, say A_2, in M_2 to collide with a message, say A_1, in M_1, A_2 must be able to "catch up with" A_1 on the downward bus at column y. To prove that message collisions cannot happen, it suffices to show that no message in M_2 will be able to catch up with any message in M_1. Or equivalently, A_1 has been switched out of the downward bus before A_2 catches up with it. We will prove in Proposition 2 that the condition in the Lemma that the two messages are passing processor (x, y) on the same bus does not hold. More formally, in the first quadrant, where $0 \le x, y < n/2$, if $S_{DR}(x, y) = (\lambda_1, 1)$ (switch control for M_1) and $S_{LD}(x, y) = (\lambda_2, 1)$ (switch control for M_2), it will be proven that $\lambda_1 < \lambda_2$.

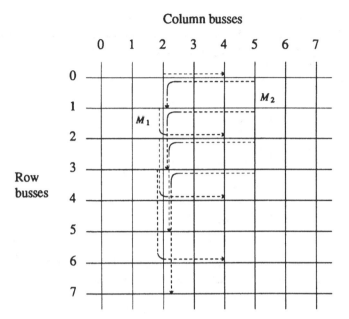

Figure 6. Potential interference between messages in M_1 and M_2.

Proposition 2. The perfect shuffle communication using one cycle is collision-free.

Proof. First we show, using the previous definition of A_1 and A_2, that A_1 has been switched out of the downward bus in column y before A_2 catches up with it and that messages in M_2 are not switched out of their destination column y.

From Eq. (4a), we have $\lambda_1 = x/2$, where x is even. Since $x \leq n/2 - 1$ in the first quadrant, we have $x/2 \leq (n-2)/4$ if $n/2$ is odd, and $x/2 \leq (n-4)/4$ if $n/2$ is even. That is

$$\lambda_1 = \frac{x}{2} \leq \begin{cases} \dfrac{n-2}{4}, & \dfrac{n}{2} \text{ odd} \\[2mm] \dfrac{n-4}{4}, & \dfrac{n}{2} \text{ even} \end{cases}$$

Similarly it can be shown that

$$\lambda_2 = \frac{n-y}{2} \geq \begin{cases} \dfrac{n+2}{4}, & \dfrac{n}{2} \text{ odd} \\[2mm] \dfrac{n+4}{4}, & \dfrac{n}{2} \text{ even} \end{cases}$$

Let $\Delta\lambda_{2,1}$ denote the delay, in number of petit cycles, between the times at which messages from M_1 and M_2, respectively, arrive at column y. Then

$$\Delta\lambda_{2,1} = \lambda_2 - \lambda_1 \geq 1$$

Therefore the potential colliding messages in M_1 have been switched out of the downward bus at column y before they collide with those in M_2.

Since any message in M_2 will arrive at column y at least one petit cycle after any message in M_1 is switched out of that column, by allowing $S_{DR}(x,y)$ to remain at *cross* for one petit cycle, i.e., by setting $\mu = 1$, the switch will be set to *straight* again, thus reestablishing the path on the downward bus at column y for messages in M_2 when they arrive. Therefore no message in M_2 is switched out of the downward bus at column y.

Similarly it can be shown that in the fourth quadrant, messages in M_3 and M_4 will not collide and no message in M_3 will be switched out of its destination column.

Next we show that in the second quadrant (the case for the third quadrant is similar), where situation is different from that in the first or fourth quadrant, no two messages A_2 and A_4, from sets M_2 and M_4, respectively, will collide. In the mean time, A_4 should not be switched out of the leftward bus in their destination row x.

Let $S_{LD}(x,y) = (\lambda_2, 1)$ (switch control for M_2) and $S_{UL}(x,y) = (\lambda_4, 1)$ (switch control for M_4). Then it can be shown that

$$\lambda_2 = \frac{n-y}{2} \leq \begin{cases} \dfrac{n-2}{4}, & \dfrac{n}{2} \text{ odd} \\[2mm] \dfrac{n}{4}, & \dfrac{n}{2} \text{ even} \end{cases}$$

and

$$\lambda_4 = \frac{n-x-1}{2} \geq \begin{cases} \dfrac{n+2}{4}, & \dfrac{n}{2} \text{ odd} \\[2mm] \dfrac{n}{4}, & \dfrac{n}{2} \text{ even} \end{cases}$$

Thus

$$\Delta\lambda_{4,2} = \lambda_4 - \lambda_2 \geq 0$$

That is, at least one message, say A_4, in M_4 will arrive on the leftward row bus at some processor (x_1, y_1) at the same time as a message, say A_2, in M_2. Will collision occur? No, since our switch settings can successfully split the two message A_2 and A_4 into different busses at processor (x_1, y_1). Figure 7 shows such a case, where A_2 and $\hat{A}_4$, arriving at (x_1, y_1) simultaneously at time $(n - y_1)/2$ (which is the time at which $S_{LD}(x_1, y_1)$ is set to *cross*), are split into the downward and leftward bus, respectively. (Our routing approach has such capability of splitting simultaneously arriving messages at a processor because we deliberately chose to transmit messages in counterclockwise direction. In other words, if the transmission had been clockwise, message collisions would have occurred at processor (x_1, y_1) in Figure 7.) It is obvious from Figure 7 that A_4 is not switched out of row x_1, which is its destination row, at processor (x_1, y_1).

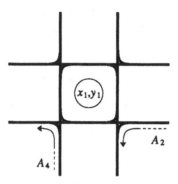

Figure 7. Two messages A_2 and A_4 arriving at processor (x_1, y_1) simultaneously are split into the downward and leftward bus, respectively, thus avoiding message collision.

In fact A_4 will not be switched out of its destination row x_1 at any other processor (x_1, y_2), where $y_2 < y_1$ and y_2 is even. To show this, note that since $S_{LU}(x, y) = (0, 0)$ for all x and y (that is, no message will be switched in leftward-to-upward direction), A_4 may only be switched out of row x_1, if possible, by $S_{LD}(x_1, y_2)$. In the following we show that by the time A_4 arrives at processor (x_1, y_2), switch $S_{LD}(x_1, y_2)$ is at the *straight* state. Specifically, if t_4 is the time at which A_4 arrives at processor (x_1, y_2) on the leftward bus, and t_2 is the time at which $S_{LD}(x_1, y_2)$ is set back to the *straight* state (after being at *cross* for one petit cycle), then we want to show that $t_4 - t_2 \geq 0$.

As mentioned previously, A_4 arrives at processor (x_1, y_1) at time $(n - y_1)/2$, which is equal to the time at which $S_{LD}(x_1, y_1)$ should be set to cross. After an additional time $y_1 - y_2$, A_4 will reach processor (x_1, y_2) on the leftward bus. Thus $t_4 = (n - y_1)/2 + y_1 - y_2$. Since $S_{LD}(x_1, y_2)$ is set to *cross* at time $(n - y_2)/2$, and remains at *cross* for one petit cycle, then $t_2 = (n - y_2)/2 + 1$. Therefore

$$t_4 - t_2 = (\frac{n - y_1}{2} + y_1 - y_2) - (\frac{n - y_2}{2} + 1) = \frac{y_1 - y_2 - 2}{2}$$

Since both y_1 and y_2 are even (see the condition in Eq. 4b) and $y_1 > y_2$, we have $y_1 \geq y_2 + 2$. That is,

$$t_4 - t_2 \geq 0$$

This completes the proof of Proposition 2. $\square$

This Proposition tells us that, using switching mode 3) in APPB with switches, the perfect shuffle communication can be performed in one bus cycle without message collision.

5. Concluding Remarks

We have presented two approaches to performing the perfect shuffle communication on processor arrays interconnected with row and column optical busses, where optical switches are used to switch optical signals directly between row and column busses. The first approach uses two bus cycles, while the second approach uses one bus cycles but requires more complex switch controls and is thus more difficult to implement. Both approaches are shown to be collision-free.

These approaches can be used for emulating the shuffle-exchange network of size n^2 on an APPB with switches of size $n \times n$. Specifically, assume n is even, then the exchange communication is simply the communication between two adjacent processors on a row bus. Combining it with the perfect shuffle communication, we can emulate the shuffle-exchange network on APPB with switches such that any two neighboring processors in the shuffle-exchange network can communicate with each other using one or two bus cycles. Such emulation allows any DSP algorithms designed for the shuffle-exchange network to be efficiently executed on APPB with switches.

References

1. R.C. Alferness, L.L. Buhl, S.K. Korotky, and R.S. Tucker, "High-Speed $\Delta\beta$-Reversal Directional Coupler Switch," *Topical Meeting on Photonic Switching, Technical Digest Series*, vol. 13, pp. 77-78, 1987.

2. K.H. Brenner and A. Huang, "Optical Implementations of the Perfect Shuffle Interconnections," *Applied Optics*, vol. 27, no. 1, pp. 135-137, 1988.

3. D.M. Chiarulli, R.G. Melhem, and S.P. Levitan, "Using Coincident Optical Pulses for Parallel Memory Addressing," *IEEE Computer*, pp. 48-57, 1987.

4. R. Cypher, J.L.C. Sanz, and L. Snyder, "Hypercube and Shuffle-Exchange Algorithms for Image Component Labeling," *J. Algorithms*, vol. 10, pp. 140-150, 1989.

5. J.R. Erickson and H.S. Hinton, "Implementing a Ti:LiNbO$_3$ 4×4 Nonblocking Interconnection Network," *SPIE Integrated Optical Circuit Engineering*, vol. 578, pp. 201-206, 1985.

6. Z. Guo, "Array Processors with Pipelined Busses and Their Implication in Optically and Electronically Interconnected Multiprocessor Architectures," Ph.D. Thesis, Department of Electrical Engineering, University of Pittsburgh, 1991.

7. Z. Guo, R.G. Melhem, R.W. Hall, D.M. Chiarulli, and S.P. Levitan, "Array Processors with Pipelined Optical Busses," *Proc. 3rd Symp. on Frontiers of Massively Parallel Computation*, pp. 333-342, 1990.

8. Z. Guo, R.G. Melhem, R.W. Hall, D.M. Chiarulli, and S.P. Levitan, "Pipelined Communications in Optically Interconnected Arrays," *J. Parallel and Distributed Computing*, to be published.

9. B.S. Kawasaki, K.O. Hill, and R.G. Lamont, "Biconical-Taper Single-Mode Fiber Coupler," *Optics Letters*, vol. 6, no. 7, pp. 327-328, 1981.

10. S.P. Levitan, D.M. Chiarulli, and R.G. Melhem, "Coincident Pulse Techniques for Multiprocessor Interconnection Structures," *Applied Optics*, vol. 29, no. 14, pp. 2024-2033, 1990.

11. A.W. Lohmann, W. Stork, and G. Stucke, "Optical Perfect Shuffle," *Applied Optics*, vol. 25, no. 10, p. 1530, 1986.

12. R.N. Mahapatra, V. Ashok Kumar, B.K. Das, and B.N. Chatterji, "Performance of Parallel FFT Algorithm on Multiprocessors," *International Conf. on Parallel processing*, vol. III, pp. 368-369, 1990.

13. R.G. Melhem, D.M. Chiarulli, and S.P. Levitan, "Space Multiplexing of Waveguides in Optically Interconnected Multiprocessor Systems," *Computer J.*, vol. 32, no. 4, pp. 362-369, 1989.

14. M.C. Pease, "An Adaption of the Fast Fourier Transform for Parallel Processing," *J. ACM*, vol. 15, no. 2, pp. 252-264, 1968.

15. C.S. Raghavendra and V.K. Prasanna-Kumar, "Permutations on Illiac-IV Type Networks," *IEEE Trans. Comput.*, vol. C-37, no. 7, pp. 662-669, 1986.

16. D.B. Sarrazin, H.F. Jordan, and V.P. Heuring, "Digital Fiber-Optic Delay Line Memory," *SPIE Proc., Digital Optical Computing II*, vol. 1215, pp. 366-375, 1990.

17. R.V. Schmidt and R.C. Alferness, "Directional Coupler Switches, Modulators, and Filters Using Alternating $\Delta\beta$ Techniques," *IEEE Trans. Circuits and Systems*, vol. CAS-26, no. 12, pp. 1099-1108, 1979.

18. H.S. Stone, "Parallel Processing with the Perfect Shuffle," *IEEE Trans. Comput.*, vol. C-20, no. 2, pp. 153-161, 1971.

19. Q.F. Stout and B. Wagar, "Intensive Hypercube Communication. Prearranged Communication in Link-Bound Machines," *J. Parallel and Distributed Computing*, vol. 10, no. 2, pp. 167-181, 1990.

20. M.S. Whalen and T.H. Wood, "Effectively Nonreciprocal Evanescent-Wave Optical-Fibre Directional Coupler," *Electronics Letters*, vol. 21, no. 5, pp. 175-176, 1985.

21. C.L. Wu and T.Y. Feng, "The Universality of the Shuffle-Exchange Network," *IEEE Trans. Comput.*, vol. 30, no. 5, pp. 324-332, 1981.

22. A.B. Yadlowsky, "The Mock Counter: A Hybrid Optical-Electronic Counter Using Fiber Delay Line Memory," OCS Technical Report 88-04, University of Colorado at Boulder, 1988.

23. T. Yasui and H. Goto, "Overview of Optical Switching Technologies in Japan," *IEEE Commun.*, vol. 25, no. 5, pp. 10-15, 1987.

3

EXPERIMENTS WITH PARALLEL FAST FOURIER TRANSFORMS

George B. Adams III, Edward C. Bronson, Thomas L. Casavant[+],

Leah H. Jamieson, and Ray A. Kamin III

School of Electrical Engineering
Purdue University
West Lafayette, Indiana 47907

[+]Department of Electrical and Computer Engineering
University of Iowa
Iowa City, Iowa 52242

INTRODUCTION

VLSI technology has made feasible large-scale parallel processing systems. Challenges presented by such systems include how to design algorithms to take advantage of the substantial parallelism and how to obtain the maximum performance from potentially complex parallel machines. In this paper we examine parallel algorithms for the fast Fourier transform. We give an overview of the algorithm structure for highly parallel FFTs then present the results of detailed experimental studies of implementing FFTs on two parallel systems: the PASM partitionable SIMD/MIMD prototype and the Thinking Machine's CM-2 massively parallel SIMD machine.

The PASM system is a dynamically reconfigurable architecture designed to allow both SIMD and MIMD operation. FFTs have been implemented on the 16-processor PASM prototype. Detailed execution time measurements using specialized timing hardware were made for complete FFT algorithms and for components of SIMD, MIMD, and barrier synchronized MIMD implementations. The component measurements isolated the effects of floating-point arithmetic operations,

This work was supported in part by the National Science Foundation under Grant Numbers CCR-8809600, ECS-8800910 and by the NSF Software Engineering Research Center (SERC). Portions of the work were performed while Adams and Kamin were summer visitors at the Research Institute for Advanced Computer Science located at NASA Ames Research Center, and while Casavant was at Purdue.

interconnection network transfer operations, and program control overhead. The studies provide comparisons of the performance of the SIMD, MIMD, and barrier synchronized MIMD implementations and address issues including synchronization, movement to and from floating-point coprocessors, instruction access time, masking to enable and disable processing elements, and network setup and data transfer time.

The CM-2 is a massively parallel SIMD machine. FFTs were executed on 8192- and 32768-processor configurations of the CM-2. Key issues addressed in the experiments included processor utilization, memory usage, data transfer time as a function of message length, and the ratio of the FFT size to the number of processors used. The algorithms were designed for a supercomputing environment and were compared to the best existing FFT algorithms on the Cray-2, developed at NASA Ames Research Center.

In the following sections we present an overview of the algorithm structure for highly parallel FFTs, then present the experimental results from the PASM and CM-2 studies. We conclude by discussing the issues raised by the two studies.

FFT ALGORITHMS

The discrete Fourier transform *(DFT)* of an N-point sequence $\{s(n)\}$, $0 \le n < N$, is defined as:

$$S(k) = \sum_{n=0}^{N-1} s(n)e^{-j(2\pi/N)nk}, \quad 0 \le k < N.$$

Fast Fourier transform algorithms allow computation of the DFT in $O(N\log_2 N)$ serial operations. One formulation, the radix two decimation-in-frequency *(DIF)* algorithm [5, 19], divides $\{s(n)\}$ into sequences $\{s_1(n)\}$ equal to the first half of $\{s(n)\}$ and $\{s_2(n)\}$ equal to the second half of $\{s(n)\}$. The DFT of the N-point sequence can be computed in terms of the two $N/2$-point DFTs of the sequences $\{s_1(n)+s_2(n)\}$ and $\{[s_1(n)-s_2(n)]W_N^n\}$, where $0 \le n < N/2$, $W_N = e^{-j(2\pi/N)}$, and $j=\sqrt{-1}$. For N a power of two, this process is repeated $\log_2 N$ times. (If N is not a power of two, there exist techniques based on zero-padding or factorization that allow use of the radix two algorithms. Throughout the remainder of this paper we will assume that N is a power of two.) Fig. 1 shows a flow graph of the computations performed in an 8-point DIF-FFT. The algorithm consists of $\log_2 N$ stages; each stage consists of $N/2$ independent DIF-butterfly operations, shown in Fig. 2a. In general, for complex-valued inputs the butterfly entails one complex multiplication and two complex additions, corresponding to four real multiplications and six real additions. In the butterflies where the "twiddle factor" W_N^k is equal to ± 1 or $\pm j$, the multiplication can be eliminated at the expense of identifying these butterflies.

A similar derivation yields the radix two decimation-in-time *(DIT)* algorithm [5, 19], in which the N-point input sequence $\{s(n)\}$ is divided into two $N/2$-point subsequences $\{s_{even}(n)\}$ and $\{s_{odd}(n)\}$, corresponding to the even- and odd-indexed

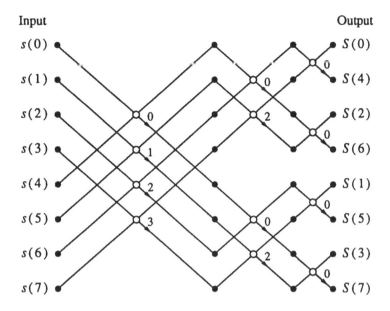

Fig. 1. Flow graph of an 8-point DIF FFT.

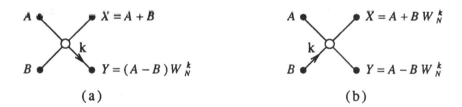

Fig. 2. Butterfly operations for (a) DIF and (b) DIT FFTs.

points in $\{s(n)\}$. The DFT of the sequence $\{s(n)\}$ can be computed as $S(k) = S_{even}(k) + W_N^k S_{odd}(k)$ for $0 \le k < N/2$ and as $S(k+n/2) = S_{even}(k) - W_N^k S_{odd}(k)$ for $N/2 \le k < N$, where S_{even} and S_{odd} are the $N/2$-point DFTs of $\{s_{even}(n)\}$ and $\{s_{odd}(n)\}$ respectively. As in the DIT algorithm, this process is repeated $\log_2 N$ times. Fig. 3 shows the flowgraph for an 8-point DIT-FFT algorithm in terms of the DIT-butterfly, which is shown in Fig. 2b. The number of arithmetic operations is the same as for the DIF-FFT.

Numerous formulations of parallel FFT algorithms have been presented, e.g., [4, 10, 11, 18, 23]. The simplest case uses $P = N/2$ processing elements *(PEs)* to compute an N-point FFT. In the most straightforward implementation, the $N/2$ butterflies at a given stage are performed simultaneously by the $N/2$ PEs. After each stage, butterfly outputs must be communicated to the appropriate PEs for the next

Input   Output

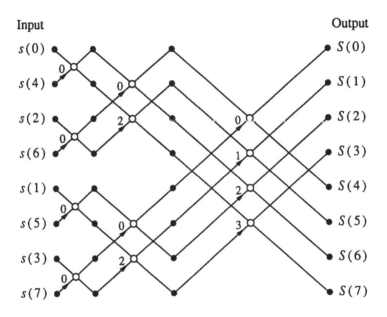

Fig. 3. Flow graph of an 8-point DIT FFT.

stage of computations. By proper selection of which PEs perform which of the butterflies at each stage, the needed data transfers can be limited to instances of the *cube* interconnection function, defined as

$$cube_c(p_{n-1}...p_{c+1}p_c p_{c-1}...p_0) = p_{n-1}...p_{c+1}\bar{p}_c p_{c-1}...p_0$$

for $0 \le c < \log_2 P$, where P is the number of PEs, $p_{n-1}...p_0$ is the binary representation of the PE's index or number, and $\bar{p}_c$ is the complement of p_c [20]. Thus, $cube_c$ specifies the connection between pairs of processors whose indices differ only in the cth bit position. The cube functions are the connections found in hypercube networks such as those in the Thinking Machines CM-2, Intel iPSC, and NCUBE machines, and form the basis for multistage networks such as those found in PASM, the BBN Butterfly, and the Ultracomputer. The computation of an 8-point DIT-FFT on 4 processors is shown in Fig. 4. The $P = N/2$-PE algorithm has a complexity of $\log_2 N$ arithmetic steps (butterfly steps) and $\log_2 N - 1$ data transfer steps. The algorithm achieves an optimal speedup on arithmetic operations and it can be shown that it achieves the lower bound on the number of data transfer steps needed when the N input data points are assumed to be distributed evenly across the $N/2$ PEs. The algorithm approach can be easily generalized to the case where $P < N/2$ by distributing the $N/2$ butterflies at each stage among the P PEs, so that each PE performs $(N/2)/P$ butterflies at each stage. In the general case, the complexity will be $(N/2P)\log_2 N$ complex multiplication steps, $(N/P)\log_2 N$ complex addition steps, and $(N/2P)\log_2 P$ data transfer steps [10].

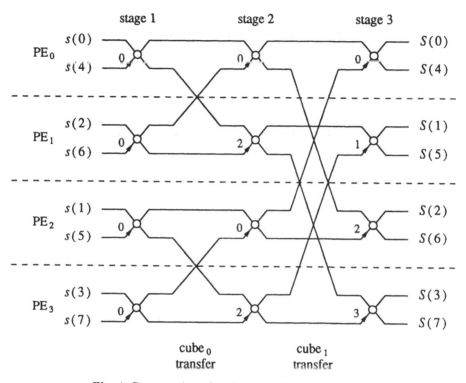

Fig. 4. Computation of an 8-point DIT-FFT on 4 PEs.

The use of $N/2$ PEs to implement an N-point FFT yields a "natural" mapping because of the $N/2$ independent butterflies computed at each FFT stage. It is also possible to use N PEs for the N-point FFT, by assigning the X and Y computations (see Fig. 2) to different PEs [12]. In this case, each PE initially holds one data point. Pairs of PEs will exchange data so that each member of the pair has the needed A and B butterfly inputs. After completion of the X and Y computations in all of the PEs, data exchanges align the inputs for the next stage of butterflies. As in the $N/2$-PE algorithm, the needed transfers can be limited to the *cube* interconnection functions. The actual time to perform the arithmetic steps in the N-PE implementation will depend on the extent to which the X and Y computations can be performed concurrently. (This will be discussed in more detail in the description of the CM-2 experiments.) The N-PE algorithm will incur an overhead of $\log_2 N$ transfer steps, compared to $\log_2 N - 1$ for the $N/2$-PE algorithm.

A common technique for reducing the arithmetic complexity of serial FFT algorithms is to use higher radix algorithms. In general, a radix-r algorithm for a 1-dimensional FFT is obtained by recursively dividing the input sequence into subsequences of length r, and requires N/r butterfly operations at each of the $\log_r N$ stages. The savings does not necessarily carry over to parallel implementations. If each processor of an SIMD machine with $N/2$ processors performs one radix-2 butterfly operation, then the total number of parallel arithmetic operations performed equals $10\log_2 N$. Similarly, a parallel radix-4 algorithm using $N/4$ processors would require $34\log_4 N$ or $17\log_2 N$ parallel arithmetic operations. Consequently, when $N/2$ processors are available fewer operations are required to use the radix-2 algorithm since $10\log_2 N < 17\log_2 N$.

In the following sections we describe the results of two detailed implementation studies. Both examine ways in which aspects of the implementation and of the target architecture affect the performance of the basic parallel FFT algorithm.

EXPERIMENTS ON PASM

Overview of the PASM Architecture

PASM is a reconfigurable architecture in which the processors can be dynamically partitioned to form independent virtual SIMD and/or MIMD machines of various sizes. This section discusses relevant characteristics of the PASM architecture. Further details can be found in [21] and [22].

The PASM Parallel Computation Unit contains P PEs (numbered from 0 to $P-1$) and an interconnection network. Each PE is a processor-memory pair. PE memory is used by the PE CPU for data storage in SIMD mode and both data and instruction storage in MIMD mode. The *Micro Controllers (MCs)* are a set of Q microprocessors that act as the control units for the PEs in SIMD mode and orchestrate the activities of the PEs in MIMD mode. Each MC controls P/Q PEs. A set of MCs and their associated PEs form a virtual machine. In SIMD mode, each MC fetches instructions and common data from its memory, executes the control instructions (e.g., branches), and broadcasts the data processing instructions to its PEs. In MIMD mode, each MC may coordinate its PEs using instructions and data from its memory.

The experiments described here were performed on a 4-MC 16-PE prototype of PASM. The PE and MC CPUs are Motorola MC68000 microprocessors operating at 8 MHz. Each PE contains a Motorola MC68881 floating-point coprocessor [15], operating at 8 or 16 MHz. Communication with the coprocessor in the PE during floating-point operations proceeds as with any peripheral.

Each PE contains special purpose hardware timing circuitry. A Motorola MC68230 timer, enhanced with additional TTL counting logic to improve resolution, was used to count processor clock cycles with an accuracy of $\pm$ 125 nanoseconds. The timer can be started or stopped by writing to a timer control register.

A simplified diagram of the structure of an MC is shown in Fig. 5. The MC contains memory from which the MC CPU reads instructions and data. Whenever the MC needs to broadcast SIMD instructions to its associated PEs, it first sets the Mask Register in the Fetch Unit to specify which PEs will execute the instructions to

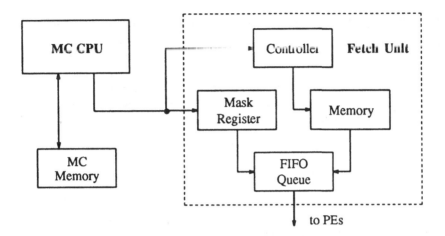

Fig. 5. Simplified MC structure with emphasis on the Fetch Unit.

follow. The MC then writes a control word to the Fetch Unit Controller to specify the location and size of a block of SIMD instructions in the Fetch Unit Memory. The Fetch Unit Controller moves this block into the Fetch Unit FIFO Queue. The current value in the Mask Register is enqueued along with each SIMD instruction.

A PE obtains SIMD instructions by performing an instruction fetch from a reserved PE memory area called the *SIMD instruction space*. Whenever logic in the PEs detects an access to this memory, a request for an SIMD instruction is sent by the PE to the Fetch Unit. When all enabled PEs have issued a request, the instruction is released by the Fetch Unit and the PEs can execute the instruction. Disabled PEs do not receive the instruction and remain idle until an instruction is broadcast for which they are enabled. Switching from MIMD to SIMD mode only requires the execution of a jump instruction from an address in memory where MIMD instructions are stored to the PE SIMD instruction space. A switch from SIMD to MIMD mode is performed by the MC broadcasting a jump instruction with a target address in the PE's MIMD space.

The SIMD instruction broadcast mechanism can also be utilized for *barrier synchronization* [13, 17] of MIMD programs. When a program requires the PEs to synchronize, the MC instructs the Fetch Unit Controller to enqueue an arbitrary data word. When the PEs executing the MIMD program need to synchronize (e.g., before a network transfer), they execute a memory read operation to the SIMD instruction space. Because the PE hardware treats SIMD instruction fetches and data reads identically, the PEs will be allowed to proceed only after the Fetch Unit has released the enqueued data and all active PEs have read it from their SIMD instruction space. This synchronizes the PEs. The same hardware that provides for SIMD synchronization among PEs also provides a fast mechanism for barrier synchronization because each barrier operation requires only a single SIMD instruction space read, regardless of the number of PEs involved in the barrier.

The MC68000 processor can execute instructions from memory with different access times. The minimum memory read or write cycle time is four clock periods. Accessing slower memory will cause the generation of one or more processor *wait states* which will increase the instruction cycle time. Each wait state requires an additional clock cycle to perform a 16-bit read or write. Each PE contains 28K bytes of 0 or 1 wait state static RAM and 2M bytes of dynamic RAM that operates at about 3 wait states. The Fetch Unit delivers data and instructions to the PEs with a delay of 2 wait states due simply to the component delays of the Fetch Unit hardware.

The interconnection network for the PASM prototype is a circuit-switched Extra-Stage Cube network [1], which is a fault-tolerant variation of the multistage cube network. In order to communicate with another PE, the initiating PE must set up a path through the network. A path is established by first writing a PE routing tag to the network *Data Transfer Register (DTR)*. The PE then sets a bit in a control register to instruct the network interface to interpret the value in the DTR as a routing tag. Byte or word data values written to the DTR will now be automatically sent through the network. The receiving PE reads the transferred data from its DTR. Upon completion, the sending PE writes to the network control register to close the path and free the network.

PASM Experiments

Implementation and Measurement Techniques. Three 4-PE parallel DIT-FFT programs, one 8-PE parallel DIT-FFT program, and two serial DIT-FFT programs were implemented and executed on PASM to study the architecture and examine the trade-offs between the different modes of parallel computation. One 4-PE 8-point SIMD program performs all FFT operations in SIMD mode. A 4-PE MIMD version calculates the butterfly operations in MIMD mode and polls the network to determine the status during network transfer operations. A 4-PE MIMD ("BMIMD" program uses barrier synchronization to align the operations of the PEs just prior to a network transfer in place of polling to test network status. The execution times for the component parts of these three 4-PE 8-point FFT programs and an additional serial 8-point FFT program were measured. An 8-PE SIMD 16-point FFT program and a serial 16-point FFT program were used to verify projected performance measures. Additional details on the PASM experiments can be found in [6].

All of the programs were written in MC68000 assembly language [14] as straight in-line code with no loops. This generated the fastest possible code and minimized issues of programming style from the architecture studies. Each program consists of an *initialization phase*, an *FFT algorithm phase*, and an *output phase*. In the initialization phase, the MC and each PE pre-compute and store all necessary data in preparation for the timing of the FFT algorithm phase. This includes ordering and initializing the input data in PE memory; pre-calculating the PE masks used by the MC; pre-calculating the logical PE number, cube function network routing tags, and FFT twiddle factors in each PE. In the FFT algorithm phase, each PE obtains the input data from PE memory, computes the FFT, and stores the transformed data back to PE memory. The execution time and the transformed data are printed during the output phase.

Execution times were measured in clock cycles using specialized PE timing hardware. Measurements were made by inserting instructions to start and stop the timers in the code before program assembly. When measuring the execution time of a complete FFT program the timers were always started and stopped simultaneously in SIMD mode. The measured times for SIMD mode operations agreed within 1 clock cycle across all PEs. In MIMD mode, since each PE timer was started and stopped independently, the measured execution times across the PEs varied. The timer overhead was measured and subtracted from the experimental program execution times.

On the PASM prototype, the origin of PE instructions influences the execution time of a parallel program. For example, measurements have shown that PE instructions will execute faster when processing in MIMD mode and when fetched from 0 wait state memory than when in processing in SIMD mode and fetched from the Fetch Unit Queue. Table 1 gives the measured PE execution times for 100 NOP instructions when fetched from various instruction stream origins.

Table 1. Average PE execution times ($\bar{x}$) and standard deviations (σ) in µs for 100 NOP instructions executed from various instruction stream origins.

Instruction Stream Origin	Execution Time		Clock Cycles per Instruction
	$\bar{x}$	σ	
dynamic RAM	89.161	0.389	7.13
Fetch Unit queue	75.003	0.021	6.00
1 wait state static RAM	62.479	0.047	5.00
0 wait state static RAM	50.000	0.000	4.00

These execution times are linearly related as can be clearly seen from Fig. 6. This variation in instruction execution time is an artifact of the prototype hardware implementation and is not a general characteristic of the architecture or mode of processing. In the experiments performed, the MIMD programs with instructions fetched from static memory enjoyed an artificial speed advantage over SIMD programs. A different hardware implementation could easily equalize the execution time of instructions fetched from different instruction stream origins. Since the focus of the study was on the relative performance of the various modes of operation, rather than the absolute execution times, the memory access time for all instructions was normalized to 2 wait states (six clock cycles). This is equivalent to the PE SIMD instruction space access time and no normalization was necessary for SIMD

58

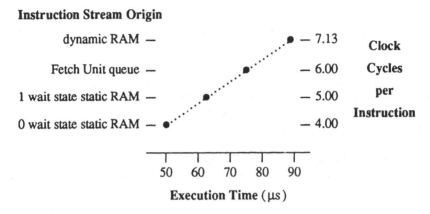

Fig. 6. PE execution times for 100 NOP instructions executed from various instruction stream origins.

instruction fetches. All other memory access cycles were normalized. During program execution, only static RAM memory was used within each PE. For SIMD mode, the static RAM was used only for data storage. To obtain the normalized execution time of any program component, the execution time was measured once using 0 wait state static RAM and again using 1 wait state static RAM. The difference between these two execution times is the time required for a single wait state per memory cycle. Adding the difference between these two execution times to the 1 wait state execution time is equivalent to the program executing from 2 wait state memory. By using this 2 wait state normalized execution time, the time of an instruction fetch in SIMD mode from the Fetch Unit Queue is equivalent to an instruction access in MIMD mode from memory. Direct comparison of program times was then possible.

Experimental Results. The execution times for the 4-PE programs are shown in Fig. 7. The serial 8-point FFT executed in 1045.000 μs. The components of the 4-PE parallel programs and the single PE 8-point FFT program were also studied. These measurements included the execution time of register initialization, program control overhead, DIT-FFT stage 1, $cube_0$ interconnection function, DIT-FFT stage 2, $cube_1$ function, and DIT-FFT stage 3 (see Fig. 4). The execution times for the components of each of the 8-point FFT programs are shown in Fig. 8. The length of each bar in Fig. 8 indicates the maximum execution time for each program component. The *FFT stage execution time* includes the time required to compute the floating-point butterfly operation plus the time required to move floating-point data to, from, and within the coprocessor. The *network execution time* is the time to transfer a complex floating-point value from the MC68000 data registers of the sending PE to the data registers of the receiving PE. This includes the time to write the routing tag to the network, request a network path, transfer the data one byte at a time, reconstruct the transferred data, and drop the network path. A solid line across

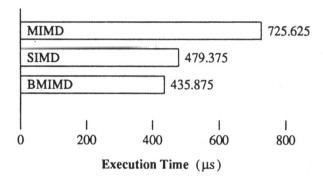

Fig. 7. Execution time for 8-point FFT programs on 4 PASM PEs.

a bar (MIMD and BMIMD Stage 3) indicates that while some of the PEs executed the program component at the maximum time indicated by the length of the bar, other PEs only required the time indicated by the solid line. This is due to the specific implementation of the FFT algorithm and will be described later. A dotted line across a bar (MIMD and BMIMD Network) indicates the minimum execution time for the program component. The measured execution times across all PEs lie between the times indicated by the length of the bar and by the dotted line.

The program components in Fig. 8 account for a minimum of 93% of the execution time of each FFT program. The remaining execution time is required for program initialization and control. Summing all of the component times for each program gives execution times of 720.688 μs for the MIMD version, 478.375 μs for the SIMD program, 432.156 μs for the BMIMD program, and 1045.000 μs for the serial version. These sums differ from the measured execution times for the complete FFT programs by -0.68%, -0.21%, -0.85%, and 0.00%, respectively.

Discussion of PASM Experimental Results

The high precision measurement techniques used in this study permitted a detailed analysis of the PASM architecture during execution of the components of each program. In this section, the differences in execution time for each of the program components are discussed.

Execution Times. Fig. 7 shows that the MIMD program has the longest execution time for any of the 4-PE parallel programs. This parallel implementation of the FFT algorithm has a speedup of 1.44 over the serial FFT program. The SIMD program requires 34% less time than the MIMD program with a speedup over the serial FFT of 2.18. The execution time of the BMIMD program is 9% less than the execution time of the SIMD program. The speedup for this program with respect to the serial program is 2.40. The reasons for the variation in execution times can be explained

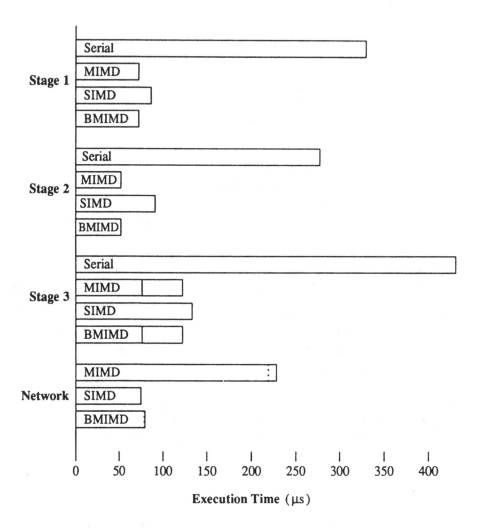

Fig. 8. Execution time for components of 8-point FFT programs on 1 and 4 PEs.

by examining the individual program components shown in Fig. 8.

In stage 1, each program executes a W_8^0 butterfly. In all modes, the butterfly requires two additions and two subtractions. The difference in the measured execution time for stage 1 is a result of the data movement from the coprocessor after the butterfly calculation. In the *cube*$_0$ network transfer that follows stage 1, PEs 0 and 2 transfer Y while PEs 1 and 3 transfer X. In MIMD and BMIMD modes, all PEs move the appropriate complex X or Y value from the coprocessor registers to the MC68000 data registers at roughly the same time. This requires two floating-point move operations. In SIMD mode, PEs 0 and 2 must first be enabled while PEs 1 and

3 are disabled and the complex Y value is moved from the coprocessor. PEs 0 and 2 are then disabled while PEs 1 and 3 are enabled and the complex X value is moved. The SIMD mode program requires two more move operations than the MIMD and BMIMD implementations.

In stage 2, the difference in measured execution time between the SIMD and MIMD programs is even greater than for stage 1. One half of the PEs perform a $W_8^0 = 1$ butterfly while the other half compute an $W_8^2 = j$ butterfly. Both of these butterfly operations require two floating-point additions and two floating-point subtractions. In the MIMD and BMIMD versions, calculation of this stage is straightforward. Each PE moves the recently transferred data values to the coprocessor registers, computes the butterfly, and moves a single complex data item from the coprocessor in preparation for the cube$_1$ network transfer. The SIMD stage 2 operation is much more complex. Although $W_8^0 = 1$ and $W_8^2 = j$ butterflies require the same number of arithmetic operations, the butterfly operations combine the A and B data values in a different order. However, it is possible to move the data into coprocessor registers in a sequence that allows the addition and subtraction operations to be performed simultaneously in all PEs. Additional masking and data movement is then necessary to prepare for the network transfer. Another reason for the longer stage 2 SIMD execution time is the necessity for an SIMD stage computation to leave the data that is not transferred in the correct floating-point registers across all PEs. This data movement is not necessary for the MIMD or BMIMD programs since each PE computes the stages independently and knows the storage locations of the data from the previous stage.

In stage 3, PEs 1 and 3 compute butterflies with non-trivial twiddle factors requiring multiply operations. PEs 0 and 2 compute the less complex W_8^0 and W_8^2 butterflies. For MIMD and BMIMD mode, the execution time for the butterflies computed by PEs 1 and 3 is indicated by the length of the bar in Fig. 8. The execution time for PEs 0 and 2 is indicated by the solid line across the bar. In SIMD mode, all of the PEs execute butterflies using multiply operations, since this is less expensive than testing for the special case twiddle factors. Since half of the PEs transferred the A value in the preceding *cube*$_1$ function and the other half transferred the B value, extra processing is required by the SIMD stage 3 to enable and disable PEs and move the data values to different coprocessor registers.

The execution time required for the interconnection network transfers varies widely among the three program implementations. The SIMD network operation requires the least amount of processing time. Since all PEs execute the network operations in lock-step fashion, the data transfers are synchronized with no need to test the network for data availability. In MIMD mode, each PE executes each butterfly independently and no implicit synchrony can be assumed when reaching the network transfer component of the program. Therefore, it is necessary for each PE to test the network (in a software polling loop) before transferring a data item and to wait on the network for a data item to become available. This testing and waiting (in software) results in high end-to-end network transfer times. Like the MIMD program, each PE executes each butterfly stage independently in the BMIMD version. However, the BMIMD version performs a barrier synchronization just prior to the network transfer. Once all of the PEs are synchronized, the data is sent and received without testing the status of the network. The execution time for the BMIMD version is slightly greater than for the SIMD version. The difference is the time required for all PEs to read from the SIMD instruction space and synchronize.

Since the execution time for the SIMD network transfer is less than the time for the barrier synchronization network transfer used in the BMIMD program, it would appear that a faster program could be constructed by using the SIMD network transfer. This is not the case. The overhead incurred by jumping to SIMD instruction space before the transfer and back to MIMD program space for the next butterfly stage exceeds the expected time savings. In addition, each time MIMD operation is resumed, it is necessary to test and branch in order for each PE to determine which butterfly operation it is to perform. The execution time overhead for these test and branch operations exceeds the time for testing and branching of an MIMD program that remains in MIMD mode and uses barrier synchronization.

In summary, the difference between the SIMD and MIMD implementations can be attributed primarily to interconnection network time. The improvement gained with the BMIMD version is principally due to MIMD execution of arithmetic operations combined with barrier synchronization prior to data transfers.

Because of the regular structure of the FFT algorithm, it is possible to derive a precise expression for the execution time of an N-point FFT program running on $N/2$ PASM PEs, $N \geq 8$. The stage times and network times measured for the 8-point FFTs can be used to construct an accurate prediction of the execution time on a larger system. A plot of the projected execution times is shown in Fig. 9. A detailed discussion of the extrapolation techniques and results can be found in [6].

The experiments on the PASM prototype clearly were not intended to demonstrate the power of the small and relatively slow prototype system. Rather, they provide detailed experimental results on the performance implications of the various synchronization protocols and explored techniques for extrapolating results to a larger system. The algorithms were hand-tailored for the highest performance that the target machine could support. In the next section, similarly highly instrumented implementations of FFTs on the massively parallel Connection Machine are described.

EXPERIMENTS ON THE CM-2

Experiments in which high-performance FFT algorithms were implemented on the CM-2 were reported in [12]. The CM-2 performance was compared to the best existing FFT algorithms on the Cray-2, developed at NASA Ames Research Center [2, 3]. The results of these experiments are summarized here.

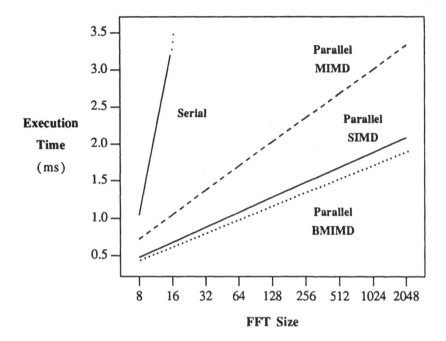

Fig. 9. Projected execution times for an N-point FFT program on 1 (Serial) and $N/2$ (Parallel) PASM PEs, $8 \leq N \leq 2048$.

Overview of the Connection Machine 2[*]

The CM-2 is an SIMD machine consisting of 64K (K=1024) PEs when fully configured [7]. The hardware with the 32-bit floating-point option consists of 2048 chip-sets where a chip-set includes two processor chips (16 PEs each), 256 KBytes of bit-addressable memory (64 Kbits per PE), a Weitek 32-bit floating point unit, and a data transposer. A 64-bit floating point option is available. Control of the Connection Machine is handled by one or more front-end host processors. The host machine provides the user with a standard operating system, I/O facilities, and debugging tools.

Physically, the 64K processors are divided into four groups of 16K PEs. Each group is controlled by a microsequencer that receives macroinstructions from the host machine, then decodes and broadcasts microinstructions to the individual PEs. These sequencers can operate independently to provide four users simultaneous access to the four quadrants of the CM-2.

---

[*]Connection Machine is a trademark of Thinking Machines, Inc.

Despite the seemingly large number of processors in the CM-2, many applications could benefit from additional processors. For this reason, the CM-2 software was designed to support *virtual processors*. The CM-2 uses time-slicing to simulate machine configurations of greater than the number of available physical processors. For example, implementing a 1,048,576 point FFT with one data point per PE on a 64K PE system requires a virtual processor ratio of 16:1. In this case, each physical processor does the work of 16 virtual processors.

The floating-point hardware consists of one Weitek 3132 floating-point accelerator and one memory interface unit for every 32 PEs. The memory interface unit is used to convert bit serial data to word parallel data and vice-versa. When a floating-point add, subtract, or multiply instruction is to be executed, each enabled PE in the group of 32 PEs sends its first operand to its memory interface chip. These values are then transferred into the accelerator chip while the second operands are loaded into the memory interface unit. Next, the second operands are transferred into the accelerator and the floating-point operation is executed. Finally, the results are sent back to the memory interface unit and subsequently returned to the appropriate PE memory. The complete cycle requires five stages. However, if a virtual processor ratio of R is used, the process is pipelined and requires only $3R + 2$ stages.

The research reported herein began on a CM-2 with 8192 processors running Version 4.3 of the system software, and continued onto a configuration of 32768 processors with floating-point hardware running Version 5.0 system software. The CM-2 and Cray-2 used are sited at the NASA Ames Research Center.

Overview of the Cray-2

The Cray-2 is a tightly coupled MIMD machine. A foreground processor handles all operating system and I/O functions. Four high-speed vector processors share 256 MWords (4096 MBytes) of common memory interleaved into 128 banks. Each processor contains eight 64-element vector registers, 16 KWords of local memory, and both vector and scalar functional units. The vector functional units or *pipelines*, reduce the overhead of vector operations, where a vector operation is an arithmetic function applied to an entire array of data.

The high performance of the Cray-2 can be partially attributed to the fast 4.1 nanosecond clock cycle time. As a consequence of this high clock rate, a potential bottleneck exists between the CPU and main memory. However, careful attention paid to algorithm design can minimize this bottleneck.

FFTs on the CM-2

Implementations. A radix-2 DIF-FFT algorithm that operates on N complex data points was implemented in C/Paris Version 4.3. C/Paris programs use standard C language statements intermixed with Paris (*Parallel Instruction Set for the CM-2*) instructions [8]. The algorithm was the basic parallelization of the DIF formulation outlined above. The twiddle factors were generated in a way compatible with the processing capabilities, memory capacity, and network speed of the CM-2. Initially, the twiddle factors are distributed among the processors according to the calculations required in the first stage. On subsequent stages, half of the twiddle factors are permuted each to two other processors. This method depends on a fast interconnection network. The twiddle factors are computed by the front-end processor and then loaded into the appropriate PEs. This program requires N processors and $6B + 2\log_2 N + 1$ bits of memory per processor, where B is the number of bits used to represent a floating-point number. Alternative methods, such as the precomputing/storing of twiddle factors used in the PASM implementation, rely on more memory than was available on the CM-2.

The C/Paris program performs the 1-D FFT with $2N\log_2 N - 2N$ real multiplications and $3N\log_2 N - N$ real additions. Since these operations are done in parallel on N processors, the actual SIMD operation count is only $4\log_2 N - 4$ real multiplications and $6\log_2 N - 2$ real additions. The total number of complex multiplications required in the first stage of a DIF-FFT algorithm can be reduced from $N/2$ to $N/2 - 4$ by taking advantage of twiddle factors of ± 1 and $\pm j$. However, on an SIMD computer all of the complex multiplications required in the first stage are executed simultaneously if $N/2$ processors are used. Thus, it is quicker to actually perform the complex multiplications of ± 1 and $\pm j$ rather than take time to disable certain processors to avoid those multiplications. This rule holds as long as at least one non-trivial complex multiplication was performed.

After the installation of Version 5.0 of the system software, two additional C/Paris programs were written, CM_2fft and CM_1fft. CM_1fft requires N PEs for an N-point FFT while CM_2fft requires only $N/2$ PEs. In the case of CM_1fft, the butterfly operation is divided between two PEs. Because the CM-2 is an SIMD machine, this means that only half of the processors can be active at any given time. $N/2$ of the PEs will first compute the X butterfly output (see Fig. 2a), while the other $N/2$ PEs remain idle. Next, the context flag in each PE is inverted thereby enabling those PEs that were previously idle and vice-versa. The Y butterfly outputs are then computed. After a butterfly is completed, the data is realigned for the next stage of the FFT. Since the data are complex values, each stage requires one 64-bit transfer when using single precision (32-bit).

As discussed in the overview of parallel FFT algorithms, the PE utilization for the N-PE CM_1fft will be only about 50%. Therefore, it appears possible to construct a program that requires $N/2$ PEs and does not suffer any degradation in performance. CM_2fft uses $N/2$ PEs, and can therefore perform an FFT twice as large as is possible with CM_1fft. Unfortunately, for problem sizes less than this upper bound CM_2fft suffers from increased overhead because of the problem of aligning the data between stages. Specifically, $N/2$ PEs need to transfer their lower butterfly operation result, while the remaining PEs need to transfer their upper butterfly result. These two results must be sent to two different memory locations in the receiving PEs. In order to perform this permutation in just one interprocessor transfer, the data must be moved to a common source location. (The move can be done more quickly than a second interprocessor transfer.) This problem could be alleviated with the implementation of the **CM_store_1L** Paris instruction. This instruction, when implemented, will allow use of indirect addressing techniques. Currently, extra data shuffling must be done, resulting in slightly lower performance figures for **CM_2fft**, in spite of the expected higher performance.

Algorithm Fine-Tuning. Careful memory utilization can improve FFT performance on the CM-2 in two ways. First, minimizing the amount of memory used to compute the basic butterfly allows computation of a larger size FFT or else more small FFTs to fit in memory. Secondly, careful allocation of memory space can reduce the number of moves needed for relocating data within the memory of a PE.

The memory usage in the one-point-per-PE algorithm, **CM_1fft**, is minimal, consisting of space sufficient only to hold the two data values and a twiddle factor. The two-points-per-PE algorithm, **CM_2fft**, benefits from a careful design, as some memory locations can be reused, reducing total memory requirements and eliminating some data movement within the memory of a PE. Fig. 10 shows the evolution in time of the memories of two communicating PEs for one iteration in the **CM_2fft** program. In one PE, the complex inputs to the butterfly are denoted A,B; in the second PE of the pair, they are denoted C,D. V_r denotes the real part of data value V and V_i the imaginary part.

Fig. 11 shows the results of a test to determine the time required for a typical data transfer in the CM-2 as a function of message length. In Fig. 11 each PE, p, sends data to PE $p + 1$ modulo 8192, assuming 8192 physical PEs. This test illustrates an interesting anomaly regarding data transfers, namely, longer messages do not necessarily take more time to transfer. Since the pipeline through the hypercube router is 29 bits, the natural message length is 29 bits. Rather than collapse each pipeline for short messages, the router microcode copies the message into a temporary location if it is less than 29 bits long, pads the message to 29 bits, and then sends the message. This copying accounts for the slight increase in delivery time as message length grows from one bit and the sudden decrease in delivery time for a message of 29 bits. Even though the FFT programs coded in this study operate on the real and imaginary parts of the complex data separately, the data transfers are treated as complex values because the time required to send a single 64-bit message is less than the time to send two 32-bit messages.

CM-2 Performance

One of the fundamental problems researchers encounter with parallel computers is the difficulty in adequately comparing the performance of machines of different architectures [9,16]. Indeed, this is the case for the CM-2 and Cray-2. An effort was made to keep many of the variables constant. Whenever possible, the CM-2 programs operate on double precision (64-bit) floating-point numbers to maintain consistency with the Cray-2 results. However, the floating-point hardware installed in the CM-2 at NASA Ames only has 32-bit capability. This should be kept

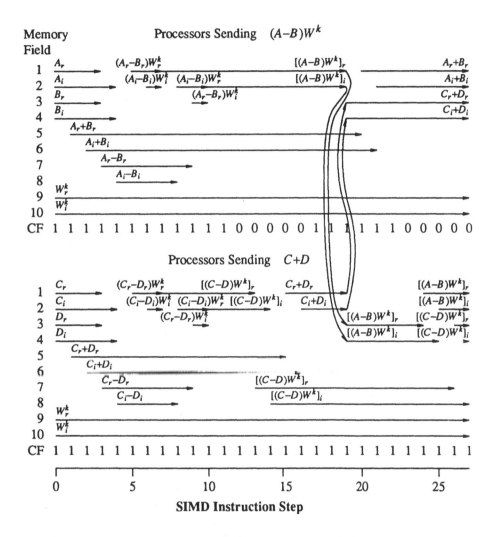

Fig. 10. PE active memory locations for **CM_2fft** as a function of the instruction being executed for a typical iteration. *CF* denotes the context flag, which determines whether a PE executes a given SIMD Instruction Step.

in mind when evaluating these results.

The FFT results are expressed in MFLOPS (millions of floating-point operations per second). Unfortunately, this unit of measure may lead to different interpretations of the results. Since it has been shown that a complex-valued radix-2 FFT can be performed using only $5N\log_2 N$ floating-point operations, this number will serve as the basis for comparisons. That is, no matter how many floating-point operations a specific complex-valued FFT program actually executes, the MFLOPS

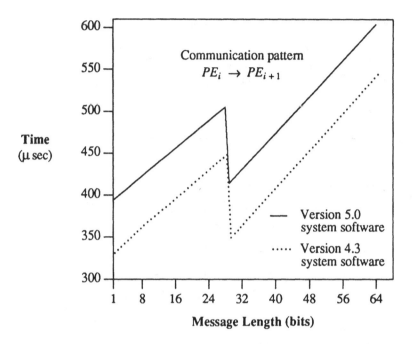

Fig. 11. CM-2 communication timings.

rate is computed using

$$rate = \frac{5N\log_2 N}{\text{execution time in } \mu \text{ seconds}} .$$

The MFLOPS rate can be misinterpreted if this guideline is not followed. Consider a program that uses more than the required $5N\log_2 N$ floating-point operations. While this program might require more actual execution time than a radix-2 version, its MFLOPS rate could potentially be higher, giving a false impression of better performance. By following the imposed guideline, both programs would be compared based on the given formula, thereby producing a higher MFLOPS rate for the radix-2 program as expected.

Measurement Techniques on the CM-2. Few tools have been developed to allow the user of the CM-2 to effectively and accurately collect performance information. A tool that can be used is **gprof**, a utility that produces an execution profile of C programs. The CM-2 Paris software includes a timing facility for reporting both real time and CM-2 active time. The timing instructions can be inserted around any block of code in a C/Paris program. Initially **CM_start_timer** clears out all the sequencer queues and makes sure all the previous Paris instructions have finished, then it reads the system time from the front end and resets the idle timer in the CM-2. The idle timer increments a counter whenever the microsequencer is waiting for an instruction from the front end. After initialization, the program executes normally until **CM_stop_timer** is encountered. At this point, the system time and idle timer are read. The elapsed front end time, CM-2 time, and CM-2 utilization can then be calculated. For accurate results the duration of time being measured must be much greater than the initialization latencies. Typically, program segments taking on the order of 10 seconds produce consistent results. Iterative loops can be used to measure the time of short code segments or even single instructions.

Experimental Results. High-performance FFT algorithm design requires a knowledge of the time to perform various elementary instructions and data transmissions. For this reason, the performance of fundamental FFT operations was measured.

In Fig. 12, the MFLOPS rate is plotted versus the virtual processor *(VP)* ratio for execution of the Paris instructions **CM_f_multiply_3_1L** and **CM_f_add_3_1L**. Both instructions read two 32-bit values from memory, perform the corresponding operation, and write the result back to a third memory location. The linear increase in performance from a VP ratio of 1:1 up to 4:1 reflects the advantage of pipelining in the 32-bit -loating point accelerator (FPA-32). Further increases in the VP ratio saturate the FPA-32 and provide a minimal increase in performance. The maximum MFLOPS rate for a 32-bit floating-point multiply and a 32-bit floating-point add were measured at approximately 2144 MFLOPS and 1867 MFLOPS, respectively.

In addition to fast arithmetic operations, high performance FFT programs also depend on rapid data movement. Fig. 13 displays the performance of the *cube$_i$* interconnection function for $0 \leq i < 16$ based on 64-bit data (the size transfers used in programs **CM_1fft** and **CM_2fft**) for VP ratios from 1 to 128. In the CM-2 16 processors are packaged in one chip, therefore, a 64K processor system has four dimensions of the hypercube on-chip and 12 dimensions connecting the chips. The data transfers can be categorized as either *on-chip* or *off-chip*. The solid line indicates the performance trend of on-chip transfers and the dotted line indicates the trend for off-chip transfers. Following the solid line indicates that the performance for *cube$_i$*, $0 \leq i \leq 3$, decreases from 13.5 millions of transfers per second to 5.2 millions of transfers per second at a VP ratio of 4:1, and then stabilizes at 5.7 millions of transfers per second for higher VP ratios. For a VP ratio of 1:1 the *cube$_i$* functions for $0 \leq i \leq 3$ are all on-chip; those for $4 \leq i \leq 15$ are all off-chip. Increasing the VP ratio to 2:1 effectively adds *cube$_4$* to the on-chip functions, and adds a virtual dimension *cube$_{16}$* to the hypercube. For a VP ratio of 128:1 *cube$_i$* functions, $0 \leq i \leq 10$, are on-chip and execute at on-chip rates.

Fig. 14 displays the MFLOPS rates of the FFT programs for a single execution of various size problems. The Cray-2 1-dimensional FFT results are based on the Cray library routine **CFFT2**. The timings are the average of 10 runs of a 64-bit FFT of various lengths. These results represent timings in a normal busy environment. If these codes were run stand-alone (with the other processors idle), the performance figures are expected to be about 30% faster. If they were run multitasked, a speedup of about four times could be expected. The Cray-2 at NASA Ames Research Center

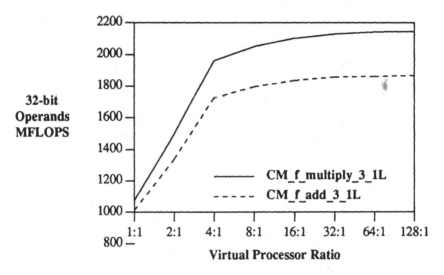

Fig. 12. MFLOPS rates for C/Paris floating-point add and multiply instructions on 32-bit operands with floating-point hardware installed.

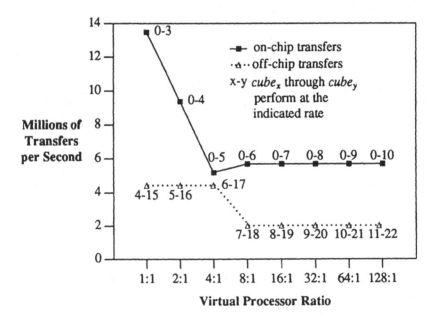

Fig. 13. CM-2 transfer rates for *cube_i* functions on 64-bit data.

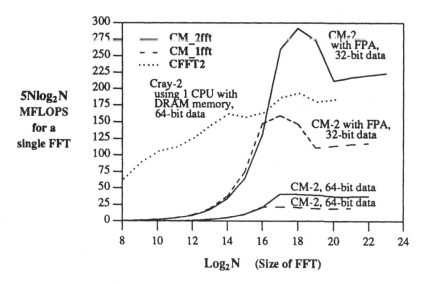

Fig. 14. $5N\log_2 N$ MFLOPS rates for one N-point FFT on a 64K-processor CM-2 and a Cray-2. CM-2 rates only assume that twiddle factors are pre-computed.

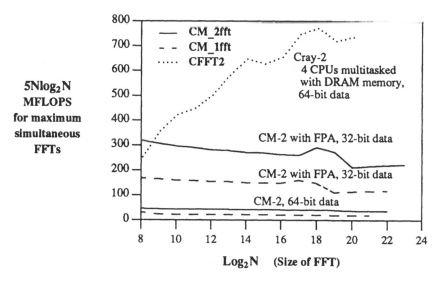

Fig. 15. $5N\log_2 N$ MFLOPS rates when computing the maximum number of simultaneous FFTs of size N on a 64K-processor CM-2, and four simultaneous size N FFTs on a 4-CPU, multitasked Cray-2. For CM-2 rates twiddle factors are pre-computed.

has 256 MWords of 80 ns DRAM memory. On a Cray-2 with SRAM memory the figures would be about 30% higher.

The March 1989 configuration of the CM-2 at Ames consists of 32K processors running at 6.7 Mhz, 32-bit floating-point hardware, three Symbolics front-ends, one VAX-8000 Series front-end, and Version 5.0 of the CM-2 system software. For the timing measurements taken, the twiddle factors, W_N^k, are assumed precomputed and available in local memory. All of the CM-2 timings were made on 32K PEs or a fraction thereof. Because a fully configured CM-2 consists of 64K PEs, the FFT results in Fig. 14 have been scaled to reflect the expected performance of a full configuration.

CM-2 memory space is the limiting factor on the maximum size of a single FFT. The program **CM_2fft** uses $10B + 49$ bits of memory per PE, where $B = 32$ for single precision and $B = 64$ for double precision. Since each PE has 65536 bits of local memory, a virtual processor ratio as high as $65536/369 = 177$ could theoretically be used for single precision data. Because the VP ratio must be a power of two, 128:1 is the maximum in this case. This implies that the largest complex-valued FFT is 8M points ($M=1048576$), assuming two points per PE, 32K physical PEs, and a VP ratio of 128:1. We can apply a similar analysis to the **CM_1fft** program, which requires $6B + 49$ bits of memory per PE. The maximum size FFT for this program is also 8M points.

The peak performance occurs at a VP ratio of 2:1 in the case of both CM-2 programs. This is the point at which the benefit of the pipelining outweighs the detriment of time-slicing. However, further increases in the VP ratio result in a decline in performance. This graph also demonstrates the advantage of the FPA-32. The single precision calculations are done by the FPA-32 while the double precision calculations are done bit-serially.

The data in Fig. 15 represents the MFLOPS rates for executing the maximum number of simultaneous FFTs. For a 64K PE CM-2 this implies that all the PEs are active. The MFLOPS rates for the Cray-2 assume all four CPUs actively compute the specified length FFT. The peak performance for the CM-2 occurs when the FFT size is small. As the size of the FFT decreases the percentage of twiddle factors that equal $W_N^0 = 1$ increases resulting in fewer required multiplies.

DISCUSSION AND CONCLUSIONS

The two studies presented were conducted on very different parallel machines, yet there is much in common in the conclusions that the studies yield. First, and most obvious, is that highly structured algorithms such as the FFT are well suited to parallel implementation. More pervasive, however, is the theme that the *highest* performance parallel FFTs are obtained at the price of a minute attention to detail in almost all aspects of the implementation. This was demonstrated in a number of instances:

- It was seen that, if the resources (i.e., number of processors) are available, the performance gains realized by higher radix serial programs do not carry over to parallel FFT implementations.

- The analytical results predicting that an $N/2$-PE algorithm would have equal or better performance than an N-PE algorithm in computing an N-point FFT were not borne out on the CM-2, due to the strict requirements for the SIMD PEs to address common locations.

- In both the SIMD PASM experiments and the CM-2 experiments, detailed coding of data movement was necessary to minimize non-uniform operations that would require enabling and disabling subsets of the PEs. In both cases, this coding yielded performance gains.

- In the PASM experiments, performance gains were obtained in going from MIMD to SIMD to barrier-MIMD implementations. As mentioned, however, some experiments, such as the MIMD program combined with SIMD transfers, yielded slower performance. The detailed experiments were needed to identify the fastest combination of processing modes.

- In the CM-2, the complex pattern of memory usage meant that careful memory utilization could both improve the data movement (and therefore execution time) and also increase the size of the FFT that could be computed.

- In the PASM study, performance gains were achieved by optimizing the use of registers in order to minimize memory references. This required attention to the specific placement of intermediate results.

- Artifacts of the network affected the packaging of data to be transferred in the CM-2, since the time required to send a single 64-bit message is less than the time to send two 32-bit messages.

- Artifacts of the hardware affected the actual (as opposed to normalized) relative execution times of the SIMD and MIMD programs on PASM, because of the different number of wait states required for execution of the SIMD and MIMD instructions.

- In both experiments, the choice of how to supply the twiddle factors for the FFT computation was dictated by the specific attributes of the machine. Alternative methods include precomputation/table look-up (used in the PASM experiments), use of the interconnection network to permute the twiddle factors to the PEs as needed (used in the CM-2 experiments), and direct calculation of the factors in each PE. The choice requires knowledge of the memory available, memory access speed, processor speed, and interconnection network speed.

Some of the effects observed, such as the variable number of wait states in PASM and the message length effects in the CM-2, have implications for hardware design. Others, such as the choice of radix-2 versus higher radix implementations based on the number of processors available and the choice of which method to use to obtain twiddle factors, require the user to make intelligent decisions based on a fairly detailed knowledge of the hardware. Both experiments performed at least some of the coding at a low level in order to make optimal use of some resource of the system. This low-level coding provides significant insights into may aspects of the architectures/systems, but places a heavy burden on the user who wishes to run "not just any parallel FFT" but "a high performance parallel FFT." The insights should be of value not only for the potential machine users, but also for the designers and builders of future parallel systems.

Acknowledgments. We thank David H. Bailey and Liviu Lustman of NASA Ames Research Center for providing FFT performance data for the Cray-2. We also thank Sam Fineberg, Wayne Nation, Pierre Pero, Tom Schwederski, and H. J. Siegel for may helpful discussions.

REFERENCES

[1] G. B. Adams and H. J. Siegel, "The Extra Stage Cube: A Fault-tolerant Inter-connection Network for Supersystems," *IEEE Trans. Comp.*, Vol. C-31, May 1982, pp. 443-454.

[2] D. H. Bailey, "A High-Performance Fast Fourier Transform Algorithm for the Cray-2" *J. Supercomputing*, Vol. 1, 1987, pp. 43-60.

[3] D. H. Bailey, "A High-Performance Fast Fourier Transform Algorithm for Vector Supercomputers" *Int'l J. Supercomputer Applications*, May 1988. 1968, pp. 275-279.

[4] G. D. Bergland, "Fast Fourier Transform Hardware Implementations -- An Overview," *IEEE Trans. Audio Electroacoust.*, Vol. AU-17, June 1969, pp. 104-108.

[5] O. E. Brigham, *The Fast Fourier Transform*, Prentice-Hall, Inc., Englewood Cliffs, New Jersey, 1974.

[6] E. C. Bronson, T. L. Casavant, and L. H. Jamieson, "Experimental Application-driven Architecture Analysis of an SIMD/MIMD Parallel Processing System," *IEEE Trans. Parallel and Distributed Systems*, Vol. 1, Apr. 1990, pp. 269-288.

[7] "The Connection Machine System Model CM-2 Technical Summary," Technical Report HA87-4, Thinking Machines Corp., Cambridge, Massachusetts, Apr., 1987.

[8] "Introduction to Programming in C/Paris," Version 5, Thinking Machines Corp., Cambridge, Massachusetts, June, 1989.

[9] P. J. Denning and G. B. Adams III, "Research Questions for Performance Analysis of Supercomputers," Research Institute for Advanced Computer

Science, RIACS Technical Report TR86.27, Dec., 1986.

[10] L. H. Jamieson, P. T. Mueller, Jr., and H. J. Siegel, "FFT Algorithms for SIMD Parallel Processing Systems," *J. Parallel and Distributed Computing*, Vol. 3, Mar. 1986, pp. 47-71.

[11] C. R. Jesshope, "The Implementation of Fast Radix 2 Transforms on Array Processors," *IEEE Trans. Computers*, Vol. C-29, Jan. 1980, pp. 20-27.

[12] R. A. Kamin III and G. B. Adams III, "Fast Fourier Transform Algorithm Design and Tradeoffs on the CM-2," *Int'l J. High Speed Computing*, Aug. 1989, pp. 207-231.

[13] S. F. Lundstrom and G. H. Barnes, "A Controllable MIMD Architecture," *1980 Int'l Conf. Parallel Processing*, Aug. 1980, pp. 165-173.

[14] Motorola, *MC68000 16/32-Bit Microprocessor Programmer's Reference Manual*, fourth edition, Prentice-Hall, Inc., Englewood Cliffs, NJ, 1984.

[15] Motorola, *MC68881 Floating-Point Coprocessor User's Manual*, first edition, MC68881UM/AD, Motorola MOS Integrated Circuits Division, Austin, Texas, 1985.

[16] D. W. Myers and G. B. Adams III, "Benchmarking and Performance Analysis of the CM-2," Research Institute for Advanced Computer Science, RIACS Technical Report TR88.19, Dec., 1988.

[17] M. O'Keefe and H. Dietz, "Hardware Barrier Synchronization: Dynamic Barrier MIMD (DBM)," *1990 Int'l Conf. Parallel Processing*, pp. I.43-I.46.

[18] M. C. Pease, "The Indirect Binary n-Cube Microprocessor Array," *IEEE Trans. Comp.*, Vol. C-26, May 1977, pp. 458-473.

[19] L. R. Rabiner and B. Gold, *Theory and Application of Digital Signal Processing*, Prentice-Hall, Inc., Englewood Cliffs, NJ, 1975.

[20] H. J. Siegel, *Interconnection Networks for Large-Scale Parallel Processing: Theory and Case Studies*, Lexington Books, D. C. Heath, Lexington, MA, 1985.

[21] H. J. Siegel et al., "PASM: a partionable SIMD/MIMD system for image processing and pattern recognition," *IEEE Trans. Comp.*, Vol. C-30, Dec. 1981, pp. 934-947.

[22] H. J. Siegel, T. Schwederski, J. T. Kuehn, and N. J. Davis IV, "An Overview of the PASM Parallel Processing System," in *Computer Architecture*, D. D. Gajski, V. M. Milutinovic, H. J. Siegel, and B. P. Furht, eds., IEEE Computer Society Press, Washington, D.C., 1987, pp. 387-407.

[23] H. S. Stone, "Parallel Processing with the Perfect Shuffle," *IEEE Trans. Comp.*, Vol. C-20, Feb. 1971, pp. 153-161.

4

Fault-Tolerance for Parallel Adaptive Beamforming

K.J.R. Liu
Electrical Engineering Dept.
Systems Research Center
University of Maryland
College Park, MD 20742

S.F. Hsieh
Dept. of Communication
Engineering
Nat'l Chiao Tung University
Hsinchu, Taiwan 30039

1 Introduction

Beamforming is used in performing spatial filtering from an array of sensors so as to minimize the undesired interferences [23]. Under nonstationary conditions, adaptive beamforming is definitely necessary if certain target performance is demanded. However, the required massive computation makes it very difficult for real-time applications. Seeking fast algorithms for adaptive beamforming is therefore of great interest for signal processing researchers. In particular, the advent of maturing VLSI technologies has triggered many researches on parallel algorithms and architectures. Among them, application specific array processors [17] are most attractive for high speed adaptive beamforming.

We will focus on two types of adaptive beamforming methods using least squares criteria, namely, sidelobe canceller (SLC) and minimum variance distortionless response (MVDR) beamformers. The recursive least squares (RLS) method is well known in its fast convergence rate, which is invariant to the eigenvalue spread of the covariance matrix of the sampled data, as compared to least mean squares (LMS) method [11,27]. In the following we will consider the SLC and MVDR beamformers using the RLS approaches.

Recent VLSI/WSI technology permits the building of million of transistors in a single chip, while a large system may require hundreds of these chips to function properly. For a complex system, a single fault from any part of the system can make the whole system useless. For various critical applications, highly-reliable computations are demanded. Fault-tolerance is therefore needed in many of these problems.

Fault-tolerance has been defined as *the ability of a system to execute specified algorithms correctly regardless of hardware failures and program errors* [42]. In order to achieve the goal of fault-tolerance, redundancy has to be introduced. When we encounter a specific VLSI signal proessing problem, an inherent nature of that signal processing algorithm can be used to develop a highly efficient specific fault-tolerant technique named *algorithm-based fault-tolerance*. The term *algorithm-based* means it is an algorithm-oriented but not a general scheme that can be applied to all general problems. A recently reported algorithm-based fault-tolerant technique, called checksum encoding (and weighted checksum) scheme proposed by Hwang, Abraham, Jou, Chen et al., has evolved from the study of VLSI matrix computation systems [31,32,33,34]. This scheme belongs to the category of information redundancy.

Since few hardware and time redundancies are necessary, it is promising for its low-cost and low overhead for VLSI/WSI multiprocessor systems. Many applications of the checksum (weighted checksum) scheme have been successfully applied to various signal processing and linear algebra operations [38]. The major drawback of the checksum scheme is that the system throughput will be slowed down because the system clock has to be extended long enough to accomodate the longer signal path of non-local interconnection caused by the checksum scheme. Unfortunately, local connection is one of the basic desirable requirements of implementations.

First, parallel algorithms and architectures fot the adaptive beamforming is briefly reviewed in the next Section. In Section 3, real-time fault-toerance for the parallel implementation of adaptive beamformers is considered. The performance analusis is given in Section 4 and the finite-precision effects are presented in Section 5. Finally, the order-degraded effects are considered in Section 6.

2 Parallel Algorithms and Architectures

2.1 RLS formulation of sidelobe canceller

A *sidelobe canceller* (SLC) comprises of a high-gain main antenna and an array of auxiliary antennas to suppress the interferences embedded in the sidelobe region of the main antenna. Like many adaptive signal processing approaches, we will adopt time recursive updating as formulations. Consider the adaptive antenna array as shown in Fig. 1 [6,11]. At each time snapshot, a $k \times p$-dimensional data A is collected by the auxiliary antennas while the main antenna receives the data $\mathbf{y}$. Our goal is to find a set of weighting coefficients w_j, $j = 1, \cdots, p$, of the beamformer and possibly its associated residual, such that the Euclidean norm of the overall residual up to time n, $\sqrt{\sum_{i=1}^{n} \|A_i \cdot \mathbf{w} - \mathbf{y}_i\|^2}$ is minimized.

If $k = 1$, Gentleman and Kung [8] in 1981 proposed a systolic array to update the optimum weight $\mathbf{w}(n)$ as time n advances by successively updating the upper-triangular matrix of the QR decomposition of the augmented matrix of $A(n)$ and $\mathbf{y}(n)$. McWhirter [21] in 1983 extended this structure by computing the most recent scalar residual at each time snapshot without explicitly computing the optimum weight vector $\mathbf{w}(n)$ which entails back substitution.

Next, recurrence formula to update the optimum weighting and residual vectors in a *block* manner as a function of time are derived. Consider a *time-recursive* least-squares (LS) problem:

$$A(n)\mathbf{w}(n) \approx \mathbf{y}(n), \tag{1}$$

where $A(n)$ and $\mathbf{y}(n)$ have growing dimensions in the number of data blocks in rows (growing-window),

$$A(n) = \begin{bmatrix} A_1 \\ \vdots \\ A_n \end{bmatrix} \in \Re^{nk \times p}, \qquad \mathbf{y}(n) = \begin{bmatrix} \mathbf{y}_1 \\ \vdots \\ \mathbf{y}_n \end{bmatrix} \in \Re^{nk}, \tag{2}$$

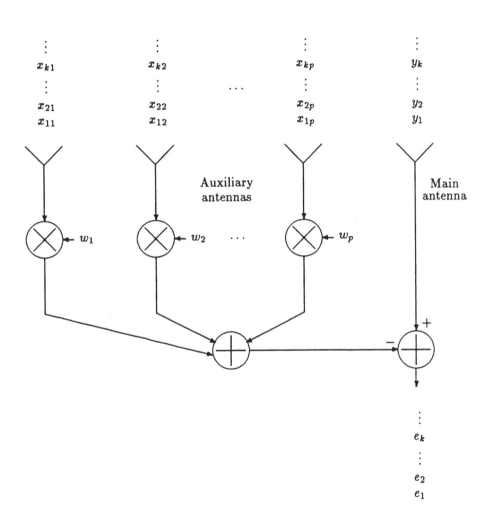

Figure 1: Adaptive antenna array

with $A_i \in \Re^{k \times p}$, $i = 1, 2, \cdots, n$ and $\mathbf{y}_i \in \Re^k$, $i = 1, 2, \cdots, n$. Here we denote k and n as the *block size* and the time index respectively, and p is the order of the LS problem. Capitalized letters (e.g., A) are used to denote matrices; small letters in boldface (e.g., $\mathbf{y}$) vectors, and small letters (e.g., w_i) scalars. A time index n is represented in the parenthesis, e.g., $A(n)$, to denote all of the time span until n, or in the subscript, e.g., A_i, the time epoch n only. For simplicity of notations, we also choose our data as real-valued. It is very easy to extend to the complex-valued cases. The previously considered scalared cases in [8,21] then have $k = 1$.

The LS solution $\mathbf{w}(n) \in \Re^p$ is computed such that the Euclidean norm of the *residual* vector

$$
e(n) = \begin{bmatrix} e_1(n) \\ e_2(n) \\ \vdots \\ e_n(n) \end{bmatrix} = A(n)\mathbf{w}(n) - \mathbf{y}(n) \in \Re^{nk} \tag{3}
$$

is minimized. All the norms $\| \cdot \|$ of vectors or matrices mentioned are 2-norm, unless otherwise specified. Our interest is to find the recurrence formula for $\mathbf{w}(n)$ and $e(n)$ as n increases.

Suppose the QR decomposition of the *augmented* matrix $[\, A(n) \ \mathbf{y}(n) \,]$ is known at time n,

$$
[\, A(n) \ \mathbf{y}(n) \,] = [\, Q(n) \ Q^{\perp}(n) \,] \begin{bmatrix} R(n) & \mathbf{u}(n) \\ 0 & \mathbf{v}(n) \end{bmatrix}, \tag{4}
$$

where $Q(n) \in \Re^{nk \times p}$ and $Q^{\perp}(n) \in \Re^{nk \times (nk-p)}$ represent the orthogonal range and null spaces of the data matrix $A(n)$, and $\mathbf{u}(n) \in \Re^p$ is the projection of $\mathbf{y}(n)$ onto $Q(n)$, $\mathbf{v}(n) \in \Re^{nk-p}$ is its counterpart projected onto $Q^{\perp}(n)$, and $R(n) \in \Re^{p \times p}$ is an upper-triangular matrix and assumed to be full-rank. $R(n)$ is sometimes called the *Cholesky factor* of the covariance matrix of $A(n)$ in that the Cholesky factorization of $A^T(n)A(n)$ can be uniquely expressed as $R^T(n)R(n)$ subject to the signs in each rows of $R(n)$ as long as $A(n)$ has full column rank.

Because an orthogonal transformation preserves the Euclidean norms of of a vector, it can be shown that[18]

$$
\begin{align}
\|e(n)\| &= \|A(n)\mathbf{w}(n) - \mathbf{y}(n)\| \tag{5} \\
&= \left\| [\, Q(n) \ Q^{\perp}(n) \,] \begin{bmatrix} R(n) & \mathbf{u}(n) \\ 0 & \mathbf{v}(n) \end{bmatrix} \begin{bmatrix} \mathbf{w}(n) \\ -1 \end{bmatrix} \right\| \tag{6} \\
&= \left\| [\, Q(n) \ Q^{\perp}(n) \,] \begin{bmatrix} R(n)\mathbf{w}(n) - \mathbf{u}(n) \\ -\mathbf{v}(n) \end{bmatrix} \right\| \tag{7} \\
&= \|Q(n)[R(n)\mathbf{w}(n) - \mathbf{u}(n)]\| + \| - Q^{\perp}(n)\mathbf{v}(n)\| \tag{8} \\
&= \| - Q^{\perp}(n)\mathbf{v}(n)\| \tag{9}
\end{align}
$$

as long as

$$
R(n)\mathbf{w}(n) = \mathbf{u}(n) . \tag{10}
$$

(8) follows from the fact that the Euclidean norm of the sum of two or more orthogonal vectors is equal to the sum of Euclidean norms of these vectors. (9) means that the residual vector while estimating $y(n)$ from $A(n)$ must lie in the null space of $A(n)$ which corresponds well with the geometrical interpretation of the orthogonal principle of LS problems.

As the time index n advances by one, i.e., a new data block of size k, $[\, A_{n+1} \; y_{n+1} \,]$, is acquired, we can write the recurrence formula for QRD as follows:

$$[\, A(n+1) \; y(n+1) \,] = \begin{bmatrix} A(n) & y(n) \\ A_{n+1} & y_{n+1} \end{bmatrix} \tag{11}$$

$$= \begin{bmatrix} Q(n) & Q^{\perp}(n) & 0 \\ 0 & 0 & I_k \end{bmatrix} \begin{bmatrix} R(n) & u(n) \\ 0 & v(n) \\ A_{n+1} & y_{n+1} \end{bmatrix} \tag{12}$$

$$= \begin{bmatrix} Q(n) & Q^{\perp}(n) & 0 \\ 0 & 0 & I_k \end{bmatrix} \begin{bmatrix} Q_{n+1} & & Q^{\perp}_{n+1} \\ & I_{nk-p} & \\ \widehat{Q}_{n+1} & & \widehat{Q}^{\perp}_{n+1} \end{bmatrix} \tag{13}$$

$$\times \begin{bmatrix} R(n+1) & u(n+1) \\ 0 & v(n) \\ 0 & v_{n+1} \end{bmatrix}$$

$$= \begin{bmatrix} Q(n)Q_{n+1} & Q^{\perp}(n) & Q(n)Q^{\perp}_{n+1} \\ \widehat{Q}_{n+1} & 0 & \widehat{Q}^{\perp}_{n+1} \end{bmatrix}$$

$$\times \begin{bmatrix} R(n+1) & u(n+1) \\ 0 & v(n) \\ 0 & v_{n+1} \end{bmatrix}$$

$$= [\, Q(n+1) \; Q^{\perp}(n+1) \,] \begin{bmatrix} R(n+1) & u(n+1) \\ 0 & v(n+1) \end{bmatrix} .$$

By defining

$$\widetilde{Q}(n+1) \equiv \begin{bmatrix} Q_{n+1} & Q^{\perp}_{n+1} \\ \widehat{Q}_{n+1} & \widehat{Q}^{\perp}_{n+1} \end{bmatrix} \in \Re^{(p+k)\times(p+k)}, \tag{14}$$

we note that $\widetilde{Q}(n+1)$ constitutes an orthogonal transformation to annihilate the newly appended data block A_{n+1}, and $Q_{n+1} \in \Re^{p\times p}$ and $\widehat{Q}_{n+1} \in \Re^{k\times p}$ represent the operation of modifying the *range* space while $Q^{\perp}_{n+1} \in \Re^{p\times k}$, and $\widehat{Q}^{\perp}_{n+1} \in \Re^{k\times k}$ that of the *null* space. We use a *hat* $\widehat{}$ to denote the new dimensional growth due to the appended data. To sum up, we have the following recurrence formula:

$$Q(n+1) = \begin{bmatrix} Q(n)Q_{n+1} \\ \widehat{Q}_{n+1} \end{bmatrix} \in \Re^{(n+1)k\times p} , \tag{15}$$

$$Q^{\perp}(n+1) = \begin{bmatrix} Q^{\perp}(n) & Q(n)Q^{\perp}_{n+1} \\ 0 & \widehat{Q}^{\perp}_{n+1} \end{bmatrix} \in \Re^{(n+1)k \times (n+1)k-p} , \tag{16}$$

$$\mathbf{v}(n+1) = \begin{bmatrix} \mathbf{v}(n) \\ \mathbf{v}_{n+1} \end{bmatrix} \in \Re^{(n+1)k-p}, \tag{17}$$

$$\begin{bmatrix} R(n+1) & \mathbf{u}(n+1) \\ 0 & \mathbf{v}_{n+1} \end{bmatrix} = \tilde{Q}(n+1) \begin{bmatrix} R(n) & \mathbf{u}(n) \\ A_{n+1} & \mathbf{y}_{n+1} \end{bmatrix}$$

$$= \begin{bmatrix} Q_{n+1} & Q_{n+1}^{\perp} \\ \hat{Q}_{n+1} & \hat{Q}_{n+1}^{\perp} \end{bmatrix} \begin{bmatrix} R(n) & \mathbf{u}(n) \\ A_{n+1} & \mathbf{y}_{n+1} \end{bmatrix}.$$

The desired optimum weighting vector $\mathbf{w}(n+1)$ and the residual vector $e(n+1)$ are thus given by

$$R(n+1)\mathbf{w}(n+1) = \mathbf{u}(n+1), \tag{18}$$

which can be solved by back substitution, and

$$e(n+1) = -Q^{\perp}(n+1)\mathbf{v}_{n+1} \qquad \text{(see (9))} \tag{19}$$

$$= -\begin{bmatrix} Q^{\perp}(n) & Q(n)Q_{n+1}^{\perp} \\ 0 & \hat{Q}_{n+1}^{\perp} \end{bmatrix} \begin{bmatrix} \mathbf{v}(n) \\ \mathbf{v}_{n+1} \end{bmatrix} \tag{20}$$

$$= \begin{bmatrix} -Q^{\perp}(n)\mathbf{v}(n) - Q(n)Q_{n+1}^{\perp}\mathbf{v}_{n+1} \\ -\hat{Q}_{n+1}^{\perp}\mathbf{v}_{n+1} \end{bmatrix} \tag{21}$$

$$= \begin{bmatrix} e(n) - Q(n)Q_{n+1}^{\perp}\mathbf{v}_{n+1} \\ -\hat{Q}_{n+1}^{\perp}\mathbf{v}_{n+1} \end{bmatrix} \in \Re^{(n+1)k}. \tag{22}$$

To see the changes of residuals in each previous data blocks due to a new observation of A_{n+1} and $\mathbf{y}_{n+1}$, we can write down the following lemma.

Lemma 1 *(updating residual)*

$$e(n+1) = \begin{bmatrix} e_1(n+1) \\ e_2(n+1) \\ \vdots \\ e_n(n+1) \\ e_{n+1}(n+1) \end{bmatrix} = \begin{bmatrix} e_1(n) - \hat{Q}_1 Q_2 \cdots Q_n Q_{n+1}^{\perp}\mathbf{v}_{n+1} \\ e_2(n) - \hat{Q}_2 Q_3 \cdots Q_n Q_{n+1}^{\perp}\mathbf{v}_{n+1} \\ \vdots \\ e_n(n) - \hat{Q}_n Q_{n+1}^{\perp}\mathbf{v}_{n+1} \\ -\hat{Q}_{n+1}^{\perp}\mathbf{v}_{n+1} \end{bmatrix} \in \Re^{(n+1)k} \tag{23}$$

proof *(23) can be derived from (15) and by noting that $Q(1) = \hat{Q}_1$, i.e.,*

$$Q(2) = \begin{bmatrix} \hat{Q}_1 Q_2 \\ \hat{Q}_2 \end{bmatrix}$$

$$Q(3) = \begin{bmatrix} Q(2)Q_3 \\ \hat{Q}_3 \end{bmatrix} = \begin{bmatrix} \hat{Q}_1 Q_2 Q_3 \\ \hat{Q}_2 Q_3 \\ \hat{Q}_3 \end{bmatrix}$$

$$\vdots$$

$$Q(n) = \begin{bmatrix} \widehat{Q}_1 Q_2 \cdots Q_n \\ \widehat{Q}_2 Q_3 \cdots Q_n \\ \vdots \\ \widehat{Q}_n \end{bmatrix} \tag{24}$$

and substituting $Q(n)$ back into (22).

(23) explains that the overall residual vector at time $n+1$ comprises of two parts: one of them is equal to $-\widehat{Q}_{n+1}^{\perp} \mathbf{v}_{n+1}$, the new dimensional growth due to A_{n+1}, while the other one is equal to the old residual vector at the previous time n, $\mathbf{e}(n)$, offset by $Q(n)Q_{n+1}^{\perp} \mathbf{v}_{n+1}$. Therefore, if we are only interested in $R(n+1)$ and/or $\mathbf{e}_{n+1}$, then we can simply maintain the information of $R(n)$ and $\mathbf{u}(n)$, which is usually the case for many applications such as beamforming[21, 26]. However, if we need to monitor all of those previously block residual vectors $\mathbf{e}_i$, $i = 1, \cdots, n$, then the previously computed range space $Q(n)$ is still required to update those old residual vectors. This monitoring may aid in the determination of some *spurious* observations(rows) such that they can be deleted (downdated) from the LS estimation problem and mitigate the possible bias caused by them [24].

2.2 Simultaneously up/down-dating RLS problems

Modifications of matrix factorizations have been of great interest in many applications [3,9,18]. In particular, recursively up/down dating QR decomposition by adding some new and deleting some old data rows will be examined, which will then lead to the systolic implementations in the following chapters.

A time-recursive up/down-dating RLS problem amounts to find $[R(n+1), \mathbf{u}(n+1)], \mathbf{v}(n+1)$ and/or $\mathbf{e}(n+1)$ from the knowledge of $[R(n), \mathbf{u}(n)]$, $\mathbf{v}(n), \mathbf{e}(n)$, the new data block $[A_{n+1}, \mathbf{y}_{n+1}]$, and the old data block $[A_{n-\ell+1}, \mathbf{y}_{n-\ell+1}]$, i.e.,

$$\begin{bmatrix} R(n) & \mathbf{u}(n) \\ 0 & \mathbf{v}(n) \\ \boxed{A_{n+1}} & \boxed{\mathbf{y}_{n+1}} \\ \boxed{A_{n-\ell+1}} & \boxed{\mathbf{y}_{n-\ell+1}} \end{bmatrix} \overset{Up/Downdating}{\Longrightarrow} \begin{bmatrix} R(n+1) & \mathbf{u}(n+1) \\ 0 & \mathbf{v}(n) \\ 0 & \overline{\mathbf{v}}_{n+1} \\ 0 & \underline{\mathbf{v}}_{n-\ell+1} \end{bmatrix} \tag{25}$$

or symbolically

$$
\begin{bmatrix}
\times & \times & \times & \cdots & \times & \times \\
 & \times & \times & \cdots & \times & \times \\
 & & \times & \cdots & \times & \times \\
 & & & \ddots & \vdots & \vdots \\
 & & & & \times & \times \\
 & & & & & \vdots \\
+ & + & + & \cdots & + & + \\
- & - & - & \cdots & - & -
\end{bmatrix}
\quad \overset{Up/Downdating}{\Longrightarrow} \quad
\begin{bmatrix}
\otimes & \otimes & \otimes & \cdots & \otimes & \otimes \\
 & \otimes & \otimes & \cdots & \otimes & \otimes \\
 & & \otimes & \cdots & \otimes & \otimes \\
 & & & \ddots & \vdots & \vdots \\
 & & & & \otimes & \otimes \\
 & & & & & \vdots \\
0 & 0 & 0 & \cdots & 0 & \oplus \\
0 & 0 & 0 & \cdots & 0 & \ominus
\end{bmatrix}.
$$

(26)

Following (4), then we have

$$
\begin{bmatrix}
A(n) & \mathbf{y}(n) \\
A_{n+1} & \mathbf{y}_{n+1} \\
A_{n-\ell+1} & \mathbf{y}_{n-\ell+1}
\end{bmatrix}
=
\begin{bmatrix}
Q(n) & Q^\perp(n) & & \\
 & & I_k & \\
 & & & I_k
\end{bmatrix}
\begin{bmatrix}
R(n) & \mathbf{y}(n) \\
0 & \mathbf{v}(n) \\
A_{n+1} & \mathbf{y}_{n+1} \\
A_{n-\ell+1} & \mathbf{y}_{n-\ell+1}
\end{bmatrix}
$$

(27)

$$
=
\begin{bmatrix}
Q(n) & Q^\perp(n) & & \\
 & & I_k & \\
 & & & I_k
\end{bmatrix}
\begin{bmatrix}
H_{11} & & H_{12} & H_{13} \\
 & I & & \\
H_{21} & & H_{22} & H_{23} \\
H_{31} & & H_{32} & H_{33}
\end{bmatrix}
$$

(28)

$$
\times
\begin{bmatrix}
R(n+1) & \mathbf{u}(n+1) \\
0 & \mathbf{v}(n) \\
0 & \overline{\mathbf{v}}_{n+1} \\
0 & \underline{\mathbf{v}}_{n-\ell+1}
\end{bmatrix},
$$

and the extended up/down-dated residual vector is given by

$$
\overline{\underline{\mathbf{e}}}(n+1) =
\begin{bmatrix}
A(n) & \mathbf{y}(n) \\
A_{n+1} & \mathbf{y}_{n+1} \\
A_{n-\ell+1} & \mathbf{y}_{n-\ell+1}
\end{bmatrix}
\begin{bmatrix}
\mathbf{w}(n+1) \\
-1
\end{bmatrix}
$$

(29)

$$
=
\begin{bmatrix}
-Q^\perp(n)\mathbf{v}(n) - Q(n)H_{12}\overline{\mathbf{v}}_{n+1} - Q(n)H_{13}\underline{\mathbf{v}}_{n-\ell+1} \\
-H_{22}\overline{\mathbf{v}}_{n+1} - H_{23}\underline{\mathbf{v}}_{n-\ell+1} \\
-H_{32}\overline{\mathbf{v}}_{n+1} - H_{33}\underline{\mathbf{v}}_{n-\ell+1}
\end{bmatrix}.
$$

(30)

If we replace the augmented LS equations in (4) by premultiplying a diagonal weighting matrix $\Lambda(n) = diag(\lambda^{n-1}I_k, \cdots, \lambda I_k, I_k)$ to diminish the importance of those previous observations (rows), where $\lambda \in (0,1]$ is a block forgetting factor, then the exponentially weighted residual in (22) will now become

$$
\Lambda(n+1)\mathbf{e}(n+1) =
\begin{bmatrix}
\lambda \mathbf{e}(n) - Q(n)Q_{n+1}^\perp \mathbf{v}_{n+1} \\
-\widehat{Q}_{n+1}^\perp \mathbf{v}_{n+1}
\end{bmatrix},
$$

(31)

where we can see that the previous residuals are gradually deemphasized by λ. Equivalently, we may consider the weight vector $\mathbf{w}(n)$ is chosen such that the

λ-weighted Euclidean norm of residual vector,

$$\|e(n)\|_\lambda = \sqrt{\sum_i^n \|\lambda^{n-i} e_i\|^2} \tag{32}$$

is minimized.

A fixed-window or sliding-window RLS filtering needs to incorporate the new data segments (updating) and also remove the influence of the obsolete data (downdating). If we denote ℓ as the number of blocks of the fixed-window size, then for $n \geq \ell$, the fixed-windowed data can be written from the growing-window data given in (2) by discarding the oldest $n - \ell$ blocks of data, i.e.,

$$A(n) = \begin{bmatrix} A_{n-\ell+1} \\ \vdots \\ A_n \end{bmatrix} \in \Re^{\ell k \times p}, \qquad y(n) = \begin{bmatrix} y_{n-\ell+1} \\ \vdots \\ y_n \end{bmatrix} \in \Re^{\ell k}. \tag{33}$$

In order to obtain $R(n + 1)$ from $R(n)$, we need to *update* (include) A_{n+1} and *downdate* (remove) $A_{n-\ell+1}$ from $R(n)$, i.e,

$$R(n + 1)^T R(n + 1) = R(n)^T R(n) + A_{n+1}^T A_{n+1} - A_{n-\ell+1}^T A_{n-\ell+1}, \tag{34}$$

where we have implicitly noticed that $R(n + 1)^T R(n + 1) = A(n + 1)^T A(n + 1)$, $R(n)^T R(n) = A(n)^T A(n)$, and $A(n+1)^T A(n+1) = A(n)^T A(n) + A_{n+1}^T A_{n+1} - A_{n-\ell+1}^T A_{n-\ell+1}$. Therefore, an updating operation in the direct data domain is equivalent to an addition in the second order domain (covariance data), while a downdating operation is equivalent to a subtraction. There are two ways to accomplish this; one is to perform updating and downdating at the same time, or we can do them one by one consecutively. These will be discussed in details in the following chapters.

Under time-varying conditions, much attention has been focused on schemes employing exponential forgetting factors, while less on fixed-windowed ones. This is partially due to the difficulty of downdating obsolete data encountered in the windowed RLS model. But, fixed-window scheme should not be precluded simply because its computational burden. Other factors, especially fast parameters tracking ability, actually favors this method under some non-stationary conditions. To motivate the need for fixed-window under non-stationary condition, a computer experiment is given to demonstrate the advantage of the faster convergence for the fixed-window method over the method based on an exponential forgetting factor.

2.3 QRD-based systolic implementations

The parallel algorithms performing recursive least squares filtering using QRD can be undertaken from the Givens rotation method, the modified Gram-Schmidt orthogonalization procedure, or the Householder transformation approach. In particular, systolic algorithms and architectures have been proposed and studied in many contexts [8,12,13,14,15,19,20,21].

Givens rotation is the most fundamental orthogonal transformation. A systolic triarray based on Givens rotations has been proposed by Gentleman and Kung [8] to perform QRD. But McWhirter [21] is the first one in tacitly propagating the rotational parameters along the diagonal direction to obtain the optimum residual without the painstaking backsubstitution for optimum weighting coefficients. Ling, et al. [19] and Kalson and Yao [15] used modified Gram-Schmidt methods to obtain similar results. A generic Givens rotation involves square-root operations which is undesirable in VLSI implementations. To avoid this difficulty, fast Givens rotation without square-roots [7,21] and another approach using CORDIC (splitting a rotation into a sequence of predetermined, multiplication-free minirotations) [1] are also of great interest. Comparisons of these algorithms such as numerical stabilities, complexities and throughput rates are still ongoing researches.

While the Givens rotation is a scalar-valued processing, modified Gram-Schmidt and Householder transformation can process vector-valued data. A systolic array using modified Gram-Schmidt methods has been reported by Hsieh and Yao [13]. Liu, et al. also proposed a systolic block Householder transformation with two-level pipelining ability [20]. It has been proved by Wilkinson [28] that the latter achieves better numerical performance compared to the more popular Givens rotations.

Although much attention has been paid on using exponentially forgetting factors, another fixed-windowed scheme using up/downdating techniques has also been studied. Rader and Steinhardt [24] first proposed hyperbolic Householder transformations. Alexander, et al. studied a hyperbolic rotation method [2]. Hsieh and Yao [14] proposed a dual-state systolic array to perform windowed RLS filtering. A RLS systolic array for the sidelobe canceller is given in Fig.2. The processing cells based on Givens rotation are given in Fig.3.

2.4   MVDR: least-squares with linear constraints

The above derived algorithms only focus on recursive LS problems with *no constraints*. If a set of *linear constraints* are incorporated, after slight modifications[20], they are still applicable. MVDR beamforming [5,22,25,26,29] is one of the examples of such applications. Suppose the data received at the antenna array is given by

$$A = \begin{bmatrix} x_1 \\ \vdots \\ x_n \end{bmatrix} \in \Re^{n \times p} , \qquad (35)$$

we want to minimize $\lambda-$norm $\| \cdot \|_\lambda$ of the residual

$$e^{(k)} \equiv A w^{(k)}, \qquad k = 1, 2, \cdots, K, \qquad (36)$$

subject to a set of independent linear constraints

$$c^{(k)^T} w^{(k)} = \mu^{(k)}, \qquad k = 1, 2, \cdots, K, \qquad (37)$$

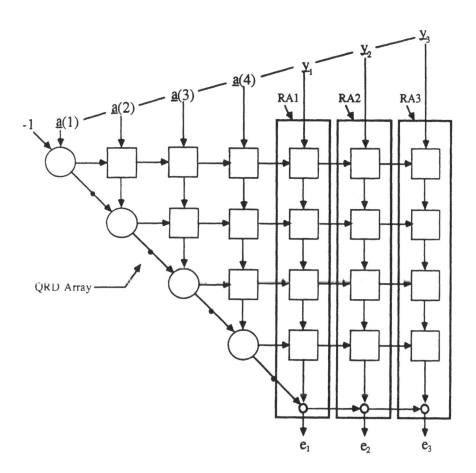

Figure 2: The RLS systolic array for sidelobe canceller

(1) Boundary Cell

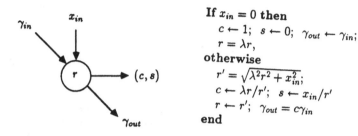

If $x_{in} = 0$ then
$\quad c \leftarrow 1; \quad s \leftarrow 0; \quad \gamma_{out} \leftarrow \gamma_{in};$
$\quad r = \lambda r,$
otherwise
$\quad r' = \sqrt{\lambda^2 r^2 + x_{in}^2};$
$\quad c \leftarrow \lambda r/r'; \quad s \leftarrow x_{in}/r'$
$\quad r \leftarrow r'; \quad \gamma_{out} = c\gamma_{in}$
end

(2) Internal Cell

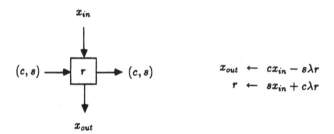

$$x_{out} \leftarrow cx_{in} - s\lambda r$$
$$r \leftarrow sx_{in} + c\lambda r$$

(3) Final Cell

$$x_{out} \leftarrow \gamma_{in} x_{in}$$

Figure 3: Processing cells based on the Givens rotation

where the λ−norm is defined as

$$\|\mathbf{v}\|_\lambda = \sqrt{\sum_{i=1}^{n} \lambda^{n-i}|\eta_i|^2}, \qquad v \in \Re^n. \tag{38}$$

The optimum weight (suppressing $^{(k)}$)is given by

$$\mathbf{w} = \frac{\mu \mathbf{R}^{-1}\mathbf{a}}{\|\mathbf{a}\|^2}, \tag{39}$$

and the corresponding residual is

$$\mathbf{e} = \frac{\mu}{\|\mathbf{a}\|^2}\mathbf{A}\mathbf{R}^{-1}\mathbf{a}, \tag{40}$$

where

$$\mathbf{a} = \mathbf{R}^{-T}\mathbf{c}, \tag{41}$$

and R is the Cholesky decomposition of the covariance matrix $\mathbf{A}^T\mathbf{A}$.

3   Real-time Fault-tolerance

3.1   Fault Model

As the VLSI technology progresses, the geometric features become smaller. Any defect affecting a given part of the circuity may cause an entire module or a logic block to become faulty and to produce arbitrary errors. Thus, the traditional gate-level single stuck-at fault model is no longer appropriate for VLSI/WSI system. A cell or module is allowed to produce arbitrary errors if any part of the cell is under failures [31]. However, we assume that at most one cell can be faulty at a given short period of time. This is based on the assumption that the system reliability is such that the mean time between failures is long enough and the probability of more than one fault occurring is very small. Some basic assumptions we need are as follows:

1. If any part of the cell become faulty, the whole cell will not function correctly;

2. The probability of the communication links and registers failing is very small and thus negligible [45].

The second assumption is reasonable since these components are typically much simpler and smaller than the processing cells themselves [45]. In addition, they can be implemented conservatively with high redundancy or with self-testing circuitry to mask a possible fault.

3.2 Concurrent Error Detection

An inherent nature of the QRD RLS systolic algorithm is that for a given data matrix A, the minimization of $\|A\underline{w}_i(n) - \underline{y}_i(n)\|_\Lambda$ for many desired response vectors $\underline{y}_i, i \in I$, can be performed concurrently by appending some more response arrays (RAs) to the systolic array. The output of the ith RA, $e_i(n)$, is the minimal residual of $\underline{u}^T(n)\underline{\hat{w}}_i(n) - d_i(n)$, where $d_i(n)$ is the nth desired response of ith input vector $\underline{y}_i$. Let $\underline{y}_0$ belong to the column space of A. That is, $\underline{y}_0 \in span\{\underline{a}(i), 1 \leq i \leq p\}$, then $\underline{y}_0 = \sum_{j=1}^p c_j\underline{a}(j)$. The optimal LS residual and the associated weight vector for $\underline{y}_0$ are $e_0(n) = 0$ and $\underline{w}_i(n) = [c_1, c_2, \cdots, c_p]^T$, for $n \geq p$. The actual selection of $\{c_1, c_2, \cdots, c_p\}$ will be given later. Various extentions of this fundamental property of the optimum residual of a LS estimation problem form the basis of the proposed *residual method* approach toward concurrent error detection. Denote $\underline{y}_0$ to be the **artificial desired response** (ADR) and the associated RA as the **error detection array** (EDA).

Lemma 2 *Given the ADR $\underline{y}_0 = \underline{a}(p)$, the contents of the EDA are identical to the contents of the pth column array of the QRD triangular array, and the optimal residual e_0, the output of the pth cell of the EDA, is always zero.*

Proof: Both arrays are $p \times 1$ vector. The first $p - 1$ elements of both arrays are identical given by the fact that the same data are rotated by same c_i and s_i generated by boundary cells $PE_{ii}, 1 \leq i \leq p - 1$, where PE_{ii} denoted the processor at position (i, i). Thus the outputs of the $(p-1)^{th}$ cells of both arrays are identical. Initially, the contents of the pth cells of both array are zeros. Let the first non-zero output of $(p-1)^{th}$ cells be x, by the update equations of both cells, the first non-zero contents of both p^{th} cells equal x. If the second non-zero output of $(p-1)^{th}$ cells is z, the updated content of the pth boundary cell equals $\sqrt{\lambda^2 x^2 + z^2}$ and that of the pth cell of the EDA is $s_p z + c_p \lambda = \sqrt{\lambda^2 x^2 + z^2}$, where $s_p = z/\sqrt{\lambda^2 x^2 + z^2}$ and $c_p = \lambda x/\sqrt{\lambda^2 x^2 + z^2}$. Therefore, the contents of both array are identical. Since the rotation coefficients, c_p and s_p, generated by PE_{pp} boundary cell are proportional to x and z respectively, the output of the pth cell of the EDA, e_0, is $c_p z - s_p x$ which is always zero. $\square$

If there is a fault in either the pth column of the array or the EDA, these contents are no longer identical and then lead to a non-zero e_0. Thus a fault is detected. However, if there is any fault outside of these two arrays, then the errors produced by that fault will affect both of these arrays in the same manner (*i.e.*, contents of both arrays are still identical) and resulting in a zero e_0. Thus, these faults will not be detected by the $\underline{y}_0 = \underline{a}(p)$ design. Clearly, we can generalize the above results by the following Lemma.

Lemma 3 *Given the ADR $\underline{y}_0 = \underline{a}(k)$, for $1 \leq k \leq p$, the contents of the kth column array of the QRD triarray and the first k cells of the EDA are identical. The output of the kth cell of the EDA are zero. The contents of cell l, $k + 1 \leq l \leq p$, of the error detection array are all zeros.*

Corollary 1 *A fault occurring outside of the kth column of the QRD triarray and the EDA will not be detected if the ADR is designed as $\underline{y}_0 = \underline{a}(k)$.*

Proof: From the previous discussion, a fault occurring in the ith, $1 \leq i \leq k-1$, column of QR array will not be detected. From *Lemma 3*, the output of the kth cell and the contents of cell l, $k + 1 \leq l \leq p$, of the EDA are all zeros. Thus, any fault occurring to the ith, $k + 1 \leq i \leq p$, column of QR array will be masked by these zeros. The optimal residual e_0 is always zerounless there is any inconsistency between the kth column of the QR array and the EDA. □

From all the above observations, by selecting $\underline{y}_0$ properly as given in the following theorem, we can detect the presence of a fault in any location of the system.

Theorem 1 (Concurrent Error Detection Theorem) *Consider the selection of the artificial desired response $\underline{y}_0 = \sum_{i=1}^p \underline{a}(i)$. If there is no fault in the system, then theoutput of the EDA with $\underline{y}_0$ as an input yields $e_0 = 0$. If there is a fault in the system, then $e_0 \neq 0$.*

Proof: From *Lemma 3*, each $\underline{a}(i)$ is "zeroed out" by the ith cell of the EDA. Any error produced by a faulty processor, say in the jth column of the QR array, will not be zeroed out by the jth cells of the EDA. The output of the jth cell is then non-zero and propagates down to the output. Therefore, whenever $e_1 \neq 0$, there is a fault in the system. □

The ADR $\underline{y}_0 = \sum_{i=1}^p \underline{a}(i)$ is obtained by implementing a top row encoding array (EA) consisting of p summing cells as shown in Fig.4. The response array (RA), with the desired response $\underline{y}$ as input and e as output, located at the right of the EDA is incorporated with the system to produce the desired residual. Once $e_0 \neq 0$, which indicates the system had a fault, then $e(n)$ is considered to be in error and will not be used. The error detection is thus achieved in real-time.

Example 1: An adaptive filter using QRD LS systolic array with order $p = 3$ is simulated. In between $t = 25$ and $t = 35$, a fault occurred in cell PE_{23} in such a way that random noise within range $[-1, 1]$ is generated. Fig.5 plots $|e_0|$ versus t and shows the adaptive effect of the algorithm. A threshold device can be used with e_0 to provide a decision on the size of the error that can be tolerated. Fig.6 shows the $|e_0|$ in Fig.5 with threshold set at 0.3. A generalization of the proposed scheme is stated below.

Theorem 2 (Generalized Error Detection Theorem) *Any fault occurring in the system can be detected if the ADR is given by $\underline{y}_0 = \sum_{i=1}^p c_i \underline{a}(i)$, where $c_i \neq 0, 1 \leq i \leq p$.* □

The simplest ADR that can detect fault is indeed a checksum encoded data (given by *Theorem 1*) which is a special case of the set of ADR given by *Theorem 2*. However, unlike the checksum fault detection scheme in [31,34], *Theorem 1 and 2* provide a real-time fault detection scheme using the inherent nature of QRD RLS systolic array.

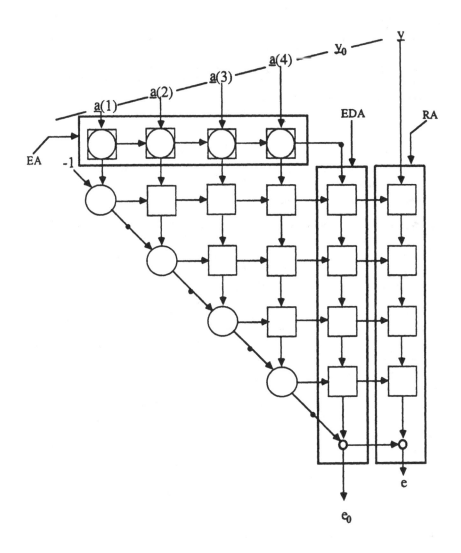

Figure 4: Fault-tolerant RLS systolic array

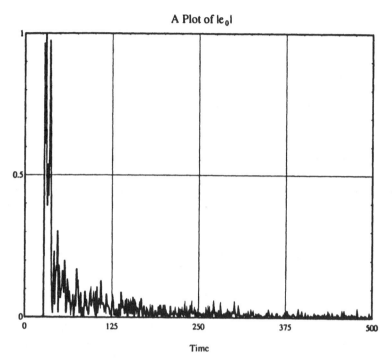

Figure 5: A plot of $|e_o|$ versus t

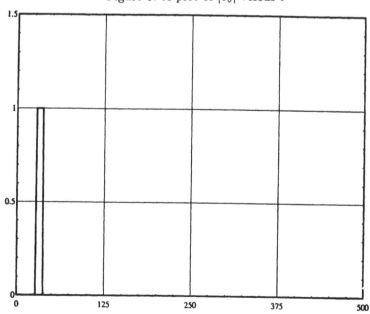

Figure 6: A plot of $|e_o|$ versus t with threshold 0.3

3.3 Fault Diagnosis

When a fault is detected, the system leaves the concurrent error detection phase and enters the fault diagnosis phase. The main purpose of this phase is to find the faulty processor row. Either of two methods, the flushing fault location (FFL) method or the checksum encoding (CSE) method, can be used to diagnose and locate the faulty row. The FFL method is developed under the assumption that only the residual output e_0 can be accessed externally, while the CSE method assumes that all the cells of the EDA can be accessed.

3.3.1 Flushing Fault Location Method

During the concurrent error detection phase, a fault is detected based on unknown values of the incoming data in $\underline{a}(k)$, $1 \leq k \leq p$, and contents of all the cells. However, in the FFL method, we will control the desired incoming data as well as the contents of the tri-array and the EDA, in order to obtain an appropriate value in e_0 to locate the faulty processor row. In the FFL method, we do not use the operations of the RA cells or the EA cells. From (10), the weight vector $\underline{\hat{w}} = (\hat{w}_1, \hat{w}_2, \cdots, \hat{w}_p)$ can be solved by using the back substitution method of

$$\hat{w}_i = \frac{P_i - \sum_{j=i+1}^{p} r_{ij} \hat{w}_j}{r_{ii}}, \quad i = p, \, p-1, \cdots, 1 \tag{42}$$

where P_i and r_{ij} are elements of vector $\underline{P}$ and matrix R. A linear array to performed the back substitution as in [16] can be prevented by using the *weight flushing technique* cite.

In a fault-free triangular LS systolic array, by "freezing" the QRD upper triangular matrix $R(n)$ and the associated column vector $\underline{P}(n)$ in (10), theoptimum solution $\underline{\hat{w}}(n)$ can be "flushed" out sequentially with a skewed identity matrix input. In these operations, the internal cells in a given row, act as pure bypass elements with no Givens rotations, when the input to the boundary cell of that row is zero and $c = 1$ and $s = 0$ are being propagated.

However, due to errors generated by the faulty processor, various parts of $R(n)$ and $\underline{P}$ stored in the array are no longercorrect. A new test triangular matrix T and a test vector $\underline{P}_t$ are loadto the tri-array and EDA respectively. The values of T and $\underline{P}_t$ can be either pre-stored or distributed by an host computer when a fault is detected. Specifically, define $T = [\underline{t}_1, \underline{t}_2, \cdots, \underline{t}_p]$ as a $p \times p$ all-1's upper triangular matrix, where $\underline{t}_i$ is a $p \times 1$ vector, and $\underline{P}_t$ as a $p \times 1$ all-1's vector. Since $\underline{P}_t = \underline{t}_p$, the optimal solution vector $\underline{\hat{w}}_0 = [0, 0, \cdots, 1]$, as given by (10). One of the reasons for the selection of these all 1's in T and $\underline{P}_t$ is to reduce memory requirement. Only one-bit register is required for each cell or distribution requirement.

With T and $\underline{P}_t$ frozen, consider a skewed $p \times p$ identity matrix input to the first p columns and all zeros input to the EDA. In the absence of a fault, components of the optimum solution vector, $\underline{\hat{w}}_0 = [0, 0, \cdots, 1]$, are outputted sequentially at e_0 [30]. Denote $\underline{\phi}_i^T = [0, \cdots, 0, 1, 0, \cdots, 0]$ as a $1 \times p$ vector of all zeros except for an one at the ith position, representing the ith row of the skewed identity matrix input to the array. Then the output $e_0(1)$ in response to $\underline{\phi}_1^T$ is processed by all the cells from the first to the pth row of the array. In general, $e_0(i)$ is the response to $\underline{\phi}_i^T$ (i.e. $e_0(i) = \underline{\phi}_i^T \underline{\hat{w}}_0$), and is processed by all the cells from the ith to the pth row. As considered above, in theabsence of a fault, $e_0(i) = 0$, $1 \leq i \leq p-1$, and $e_0(p) = 1$. However, with a fault in the kth row, then $e_0(i) \neq 0$, $1 \leq i \leq k$, but responses due to $\underline{\phi}_j^T$ for $k+1 \leq j \leq p$, encountering only fault free processing cells from the jth to the pth row, will yield the correct value. This property can be used to locate the faulty row in the FFL method.

Theorem 3 (Flushing Fault Location Theorem) *When a fault is detected and the system enters the fault diagnosis phase, both the EA and RA arrays are made inoperativeand all 1's are loaded into the contents of the processor cell. A skewed identity $p \times p$ matrix is flushed into the system with all zeros input to the EDA. The EDA output $e_0(i), 1 \leq i \leq p$ is obtained sequentially. Assume the first zero output at $e_0(k+1) = 0$, occurs for some k, $1 \leq k \leq p-2$, then the faulty processor is in the kth row. If there is no such k for $e_0(k+1) = 0, 1 \leq k \leq p-2$, and $e_0(p) = 1$, then the $(p-1)^{th}$ row is the faulty row;otherwise, with $e_0(p) \neq 1$, the pth row is the faulty row.* □

Example 2: A QRD RLS array with order $p = 5$ is considered. Suppose a fault has occurred in PE_{34}. When a skewed 5×5 identity matrix is flushed into the system, due to the randomly generated noise from the faulty cell PE_{34}, the outputs from e_0 are given by $[0.2127, -0.5714, 0.7453, 0, 1]$ In the absence of an error, the outputs should be $[0., 0., 0., 0., 1.]$. Since the first three elements are erroneous, based on *Theorem 3*, the faulty cell is in the third row.

It is obvious the flushing of $\underline{\phi}_1$ is unnecessary since computations involving the entire QR array are definitely incorrect in the fault diagnosis phase.

3.3.2 Checksum Encoding Method

The basic assumption of the CSE method is that all the cells of the EDA can be accessed. Further more, all the contents of the tri-array can be piped out in the diagnosis phase. In this paper, instead of using the CSE as a fault detector as in [31], it is used to diagnose fault location when a fault has been detected. The disadvantages of the checksum scheme for real-time application is thus prevented. It has been shown in [31] that the QRD of a row checksum matrix A_r results in a row checksum upper triangular matrix R_r. Let $r_{ij}(n)$ be the content of processor PE_{ij} of the tri-array at time n and $P_i(n)$ be the content of the ith processor of the EDA at time n.

Theorem 4 (Checksum Encoding Theorem) *Given the artificial desired response $y_0 = \sum_{i=1}^{p} \alpha_i \underline{a}(i)$, for $\alpha_i \neq 0$, the checksum*

$$\sum_{k=0}^{p-i} \alpha_i r_{i(i+k)}(n+k) = P_i(n+p-i+1), \qquad (43)$$

holds for $i = 1; 2, \cdots, p$, $n = p, p + 1, p + 2, \cdots$, if no fault has occurred. If there is an m such that the checksum does not hold for $m \leq i \leq p$, then there is a fault in the system and the faulty processor is in the first row that does not meet the checksum. $\square$

The time indexs are introduced to describe the time difference for a given row input of data to the array. Each column of input, say $\underline{a}(k)$, is zeroed out by the kth cell of EDA. Thus the content of the kth cell of EDA is affected only by $\underline{a}(i)$, $k \leq i \leq p$. Therefore, if there is a faulty processor, say in row m, then all the rows below do not satisfy the checksum because of the error produced by the faulty one which cannot be zeroed out by the mth cell of EDA.

Example 3: Consider a $[R \vdots \underline{P}_0]$ matrix of intermediate results for a QRD RLS array with order $p = 4$,

$$
\begin{bmatrix}
0.7930 & 0.7462 & 0.4655 & 0.9774 & \vdots & 2.9821 \\
0. & 1.2973 & 1.0816 & 1.0379 & \vdots & 3.4168 \\
0. & 0. & 0.2729 & 0.3643 & \vdots & 1.7132 \\
0. & 0. & 0. & 0.1675 & \vdots & 0.8375
\end{bmatrix},
$$

where the $(i, i + k)$ element, $r_{i,(i+k)}$, of R, takes the value at time $n + k$ due to time skewing of the input. That is, $r_{i,(i+k)}(n + k)$. The ith element of $\underline{P}$ takes the value at time $n + 5 - i$. That is, $P_i(n + 5 - i)$. As we can see, after the third row, the checksums are no longer met for each row. From *Theorem 4*, the faulty cell is in the third row.

Unlike the FFL method, the CSE method cannot stop the concurrent error detection phase immediately when a fault is detected. Because of the skewed manner of inputtingthe data, if we stop the operation immediately, the checksum property will not hold according to *Theorem 4*. Each processor PE_{ij}, $1 \leq i \leq j \leq p$, of the QR tri-array has to take $j - i$ more data and the ith cell of the EDA has to take $p + 1 - i$ more data so that the checksum is satisfied for each row of the systolic array. If each data requires one system clock, we observe that at most p more system clocks are needed to process those unfinished data after the moment a fault is detected. The last row takes one clock and the first row takes p clocks. Generally, the ith row takes $p + 1 - i$ clocks to process the unfinished data. Thus, those rows which take fewer clocks can pipe their final results out to the right to check their checksum while others rows are still working on their unfinished data.

3.3.3 Order-Degraded Reconfiguration

An order-degraded performance is reasonable and often acceptable in many LS applications. A reconfiguration is needed to reroute data paths for order-degraded operation. Many models and approaches can be found in the literatures [47,50,56] for the reconfiguration of VLSI array processors. Here we use a

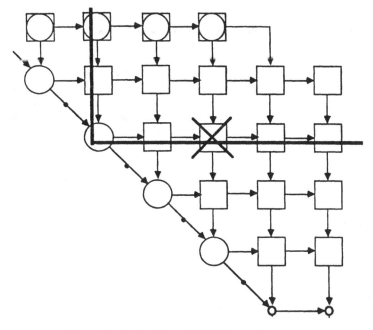

Figure 7: Order-degraded RLS systolic array

similar model described in [50]. When the faulty row, say row k, is determined, the cells in the kth column and row become connection elements and enter a dormant state. In the dormant state, each cell tests itself to check its status repeatedly [35]. The reduced $(p-1) \times (p-1)$ tri-array then operates in an order-degraded RLS computational manner. Fig.7 shows an example of bypassing the faulty row and the associated column to become an order-degraded RLS array. When the (transient) fault is removed, the dormant cells reactivate and generate an interrupt immediately. The reactivation scheme recovers all of the cells which become connection elements before and turns them into active cells. Then the full-order RLS operation is resumed. Details on various schemes and technologies of reconfiguration can be found in [57].

4 Performance Analysis

4.1 Optimal Efficiency

Since a column of linear EDA and a row of linear EA are required, the complexity of this fault detection scheme is $2p$. That is, $2p$ redundant processors is required. Compare to the complexity of the triangular QRD RLS array, $(p^2 + 3p)/2$, it is a cost-effective real-time fault detection scheme. Here we do not count the final output multiplier cells in the EDA and RA.

We define the *hardware efficiency* Ω_h to be the ratio of the hardware cost of implementing the algorithm to the cost of implementing the algorithm with an error-detection capability. We see $\Omega_h(p) = (p^2 + 3p)/(p^2 + 7p)$. Thus $1/2 \leq \Omega_h(p) \leq 1$ since a single error can be detected by duplicating the hardware. When $\Omega_h(p) = 1$, we say the error-detection scheme is *most hardware efficient*. Define the *time efficiency* Ω_t to be the ratio of the time to implement the algorithm and the time to implement the algorithm incorporating the error-detection scheme. Obviously, time efficiency is bounded by $0 \leq \Omega_t \leq 1$. When $\Omega_t = 1$, we say the error-detection scheme is *most time efficient*. If an error-detection scheme is both most hardware and time efficiency, then it is said to be *optimal*. For the proposed residual method, clearly $\lim_{p \to \infty} \Omega_h(p) = 1$. That is, it is asymptotically most hardware efficient. However, the time efficiency $\Omega_t(p) = 1, \forall p$, so that it is also most time efficient. Therefore, the residual method is an asymptotically optimal error-detection scheme for the recursive LS systolic array.

4.2 Robust Error Detection

Assume a fault occurs in an internal cell $PE_{ij}, i \neq j$, at a faulty moment. The output of this faulty cell is thus erroneous and can be described by $x_{out}^\varepsilon = x_{out} + \delta$, where x_{out} is the fault-free output and δ is the error generated by the fault. The error propagation path can be described by

$$PE_{ij} \to PE_{(i+1)j} \to \cdots \to PE_{jj},$$

and then $PE_{kl}, \ k \geq j, \ l \geq j$ are all contaminated [54]. From the operations executed by the internal cell, the error is modified to $c_{i+1}\delta$ by $PE_{(i+1)j}$ and the cumulative modifications of the error before reaching the boundary cell, PE_{jj}, is

$$\eta = \delta \prod_{k=i+1}^{j-1} c_k, \tag{44}$$

where c_i is the cosine parameter generated by the boundary cell PE_{ii}. Let c_j' and s_j' denote the erroneous c_j and s_j respectively. The c_j' and s_j' are then given by

$$c_j' = \frac{\lambda r}{\sqrt{\lambda^2 r^2 + (x_{in} + \eta)^2}}, \qquad s_j' = \frac{x_{in} + \eta}{\sqrt{\lambda^2 r^2 + (x_{in} + \eta)^2}}. \tag{45}$$

In this case, s_j' is no longer proportional to x_{in}, $\underline{a}(j)$ will not be zeroed out by the jth cell of the EDA [54]. The size of the error generated by this cell is

$$\eta_j = c_j' x_{in} - s_j' \lambda r = -\frac{\lambda r \eta}{\sqrt{r'^2 + 2\eta x_{in} + \eta^2}} = -c_j' \eta, \tag{46}$$

where $r' = \sqrt{\lambda^2 r^2 + x_{in}^2}$ is the new updated uncontaminated value of the content of PE_{jj}. When η_j propagates down to the output of the EDA, η_j is influenced by the contaminated cosines c' of each following row. The error output at e_0 due to an error δ generated at PE_{ij} is then given by

$$
\begin{aligned}
e_0^\delta(i,j) &= -\gamma \prod_{m=j+1}^{p} c'_m \eta_j = -\gamma \prod_{m=j}^{p} c'_m \eta \\
&= -\gamma \prod_{k=i+1}^{j-1} c_k \cdot \prod_{m=j}^{p} c'_m \delta.
\end{aligned}
\tag{47}
$$

where $\gamma = \prod_{l=1}^{j-1} c_l \prod_{k=j}^{p} c'_k$ [21]. It becomes

$$
e_0^\delta(i,j) = -\prod_{l=1}^{i} c_l \prod_{k=i+1}^{j-1} c_k^2 \prod_{m=j}^{p} c'^2_m \delta.
\tag{48}
$$

Next, assume a fault occurs in a boundary cell, PE_{jj}, $1 \le j \le p$, at the faulty moment. Both erroneous c'_j and s'_j produced by PE_{jj} can be written by

$$
c'_j = \frac{\lambda r + \delta_c}{r'_\epsilon}, \qquad s'_j = \frac{x_{in} + \delta_s}{r'_\epsilon},
\tag{49}
$$

where δ_c and δ_s represent errors in the numerators while r'_ϵ represents the erroneous content of the denominators of c_j and s_j. The error produced by the jth cell of the EDA is then given by

$$
\eta_j = c'_j x_{in} - s'_j \lambda r = \frac{x_{in}\delta_c - \lambda r \delta_s}{r'_\epsilon},
\tag{50}
$$

and the output error at e_0 due to a faulty boundary cell is given by

$$
\begin{aligned}
e_0^\delta(j,j) &= \gamma \prod_{m=j+1}^{p} c'_m \cdot \frac{x_{in}\delta_c - \lambda r \delta_s}{r'_\epsilon} \\
&= \prod_{l=1}^{j} c_l \cdot \prod_{m=j+1}^{p} c'^2_m \cdot \eta_j.
\end{aligned}
\tag{51}
$$

Lemma 4 *For some c_j, if there exists an $n \in I$ such that $c_j(n) \ne 0$, then $c_j(m) > 0$ for all $m > n$, $m \in I$.*
Proof: The cosine parameter is given by $c_{k+1} = \lambda r(k)/r'(k)$, where $r(k+1) = r'(k) = \sqrt{\lambda^2 r^2(k) + x_{in}^2(k)}$. $\exists n \to c_j(n) \ne 0$ is equivalent to say that $\exists n \to r(n) \ne 0$ or $r(n) > 0$. Since $r(k+1) = r'(k) \ge \lambda r(k)$, we have $r(k) > 0$ for all $k \ge n$. Therefore, $c_j(k) > 0$ for all $k > n$.$\Box$

For linearly independent input column vectors, all $c_j(n) \ne 0$, and from (48) and (51), we can see that $e_0^\delta \ne 0$, under unlimited precision condition, if there is a fault occurs in the system, except when $u_{in}\delta_c = \lambda r \delta_s$ in (50). However, this is unlikely to happen. An error produced by a faulty processor at the faulty moment will be detected at the EDA output e_0 and the probability of error detection given a fault occurs equal one. That is,

$$
Pr(\text{error detected at } e_0 \mid \text{a fault occurred}) = 1.\Box
\tag{52}
$$

4.3 Latency

Now, we consider some basic issues related to latencies in the array.

Definition 4.1: The *system latency*, t_s, is the time between the moment of data input to the system and the moment of the output of this data from the system.

Definition 4.2: The *processing latency*, t_p, of processor PE_{ij} is the time between the moment a data in a wavefront inputs to the system and the moment PE_{ij} is processing data from that wavefront.

Definition 4.3: The *error propagation latency*, t_e, isthe time between the faulty moment and the error observed moment.

It is clear that the system latency of the QR recursive LS array depends on the number of processors and delay elements on the boundary and is given by $t_s = 2p + 1$. The processing latency of processor PE_{ij}, $1 \le i \le j \le p+1$, is given by $t_p = (i+j) - 1$. Since there are a totally of $p(p+3)/2$ processors, the expected processing latency is

$$E(t_p) = \frac{2}{p(p+3)} \sum_{i=1}^{p} \sum_{j=i}^{p+1} (i+j-1) = \frac{p^2 + 4p + 1}{p+3}. \tag{53}$$

The error propogation latency is given by

$$t_e = t_s - t_p = 2(p+1) - (i+j), \tag{54}$$

and the expected value is

$$E(t_e) = (p+1)(p+2)/(p+3). \tag{55}$$

Definition 4.4: The *fault diagnosis time*, t_f, of a faulty processor PE_{ij} be the minimum time required to locate the faulty row right after the error observed moment.

Definition 4.5: The *recovery latency*, t_r, be the time between the faulty moment and the moment the faulty row is determined.

Since the system latency is $2p+1$, for the FFL method the fault diagnosis time of processor PE_{ij} for the array is $t_f^{FFL} = (2p+1) + i$. We can show that the fault diagnosis time for the CSE method is $t_f^{CSE} = p+2-i$. The expected value for fault diagnosis time are $E(t_f^{FFL}) = (5/2)p + 1$ and $E(t_f^{CSE}) = (1/2)p + 2$ respectively. By the definition of the recovery latency, we have $t_r = t_e + t_f$. Therefore, the recovery latency are $t_r^{FFL} = 4p-j+3$ and $t_r^{CSE} = 3p-2i-j+4$, while the expected recovery latency are

$$E(t_r^{FFL}) = (7p^2 + 23p + 10)/(2p+6), \tag{56}$$

$$E(t_r^{CSE}) = (3p^2 + 13p + 16)/(2p+6) \tag{57}$$

respectively. Due to the facts that multiple ports can be accessed externally and we can use the parallel pipe-out feature of the CSE method, it is not surprised that the performance of the CSE method is better than that of the

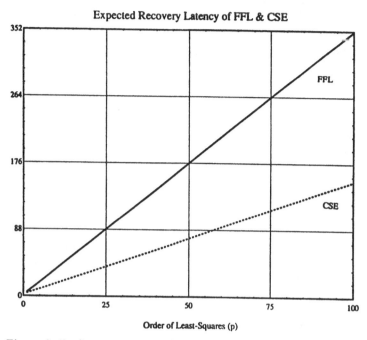

Figure 8: Performance comparisons of the CSE and FFL methods

FFL method as indicated in Fig.8. However, for both cases, the order of the expected recovery latency is $O(p)$, which is linear with respect to p. In practice, a (transient) fault may not be necessarily an immediately observable fault. Without this assumption, all the values obtained in this section become the lower bounds of those parameters. That is, performance obtained by the assumption of immediately observable fault is the best performance we can achieve.

5   Finite-Precision Effects

5.1   Missing Error Detection

The error may not be detected after multiple multiplications of c_i in (48) and (51) under finite-precision implementation. It is obvious there is no such problem when δ is large. Since r in (45) tends to be a large number asymptotically, it is reasonable to assume the error size δ generated by a fault is much smaller than r when δ is small. Under this circumstance, from (45), we have $c_j' \cong c_j$. It has been shown [55] that when λ is close to 1, the cosine parameter will eventually reach a *quasi-steady state*. In the quasi steady-state, the asymptotic behavior of erroneous cosine is $c_j' \cong c_j = \lambda$. From (48) and (51), the error output e_0^δ due to an error size δ is then approximated by

$$e_0^\delta(i,j) \cong -\lambda^{2p-i}\delta \tag{58}$$

for a faulty internal cell and

$$e_0^\delta(j,j) \cong \lambda^{2p-j}\eta_j \tag{59}$$

for a faulty boundary cell. Denote B_Δ be the wordlength of each memory and register of fixed point arithmetics. That is, each wordlength is of B_Δ bits and let $\Delta = \min(\delta, \eta_j)$. To ensure the detection of error size Δ, we need

$$\lambda^{2p-i}\Delta \geq \lambda^{2p}\Delta \geq 2^{-B_\Delta}, \tag{60}$$

Therefore, the wordlength should be at least

$$B_\Delta \geq \lceil -2p\log_2 \lambda - \log_2 \Delta \rceil \tag{61}$$

such that the small error size Δ can be detected. The second term of the right-hand size is obvious since the error size Δ must be detected; the first term is to account for the effects that the error propagates through the array of LS order p with forgetting factor λ.

5.2 False Alarm

Due to the finite-precision implementation, the residual output of the EDA will not be an actual zero if there is no fault in the system. we call this effect a false alarm. Here, we are going to model and quantitatively describe the false alarm effect and introduce a threshold device to overcome this problem.

For a QRD RLS systolic array of order p with finite-precision floating point arithmatics, denote the first row of input vector as $(\hat{x}_1, \hat{x}_2, \cdots, \hat{x}_p, \sum_{i=1}^{p}\hat{x}_i + \epsilon_p)$, where $\hat{x}_i = fl(x_i)$, $\epsilon_p = \epsilon(\sum_{i=1}^{p}\hat{x}_i)$, and $|\epsilon| < u$ is a constant[1], and the second row of input vector as $(\hat{x}'_1, \hat{x}'_2, \cdots, \hat{x}'_p, \sum_{i=1}^{p}\hat{x}'_i + \epsilon_p)$. The content of the first boundary cell is given by

$$\hat{r}_{11} = fl(\sqrt{\hat{x}_1^2 + \hat{x}_1'^2}) = \sqrt{\hat{x}_1^2 + \hat{x}_1'^2}(1+\epsilon), \tag{62}$$

and the rotation parameters are $\hat{c} = fl(\hat{x}_1/\hat{r}_{11})$ and $\hat{s} = fl(\hat{x}'_1/\hat{r}_{11})$. The contents of the internal cells can then be obtained as

$$\begin{aligned}
\hat{r}_{ij} &= fl(fl(\hat{s}\hat{x}'_j) + fl(\hat{c}\hat{x}_j)) \\
&= [\hat{s}\hat{x}'_j(1+\epsilon) + \hat{c}\hat{x}_j(1+\epsilon)](1+\epsilon) \\
&\approx (1+2\epsilon)(\hat{s}\hat{x}'_j + \hat{c}\hat{x}_j), \quad 1 < j \leq p
\end{aligned} \tag{63}$$

---

[1] To simplify the notation, we do not give indexs to different ϵ's.

and the content of the first cell of the EDA is

$$
\hat{r}_{1,p+1} = fl(fl(\hat{s}(\sum_{i=1}^{p}\hat{x}'_i + \epsilon_p)) + fl(\hat{c}(\sum_{i=1}^{p}\hat{x}_i + \epsilon_p)))
$$

$$
\approx (\hat{s}\sum_{i=1}^{p}\hat{x}'_i + \hat{c}\sum_{i=1}^{p}\hat{x}_i) + 6\epsilon_p. \tag{64}
$$

From (62), (63), and (64), the mismatch τ_1 resulted from the finite precision computation of the first row is

$$
\tau_1 = 6\epsilon_p - (\epsilon\sqrt{\hat{x}_1^2 + \hat{x}'_1^2} + 2\epsilon\sum_{i=2}^{p}(\hat{s}\hat{x}'_j + \hat{c}\hat{x}_j)) \tag{65}
$$

and it can be bounded by

$$
\begin{aligned}
|\tau_1| &\le 6p|\epsilon x_{max}| + |2\epsilon x_{max}| + 4(p-1)|\epsilon x_{max}| \\
&= (10p-2)|\epsilon x_{max}| \le 10p|\epsilon x_{max}|.
\end{aligned} \tag{66}
$$

For the second row, with the same principle, the mismatch is bounded by $10(p-1)|\epsilon x_{max}|$. The total mismatch from the whole array is given by

$$
|\tau| \le \sum_{i=0}^{p-1} 10(p-i)|\epsilon x_{max}| = 5p(p+1)|\epsilon x_{max}|. \tag{67}
$$

The possible mismatch is thus bounded by

$$
|\tau| \le 5p(p+1)|\epsilon x_{max}|. \tag{68}
$$

This bound can be interpreted as: For each row of input, each processing cell contributes about $|\epsilon x_{max}|$ amount of roundoff error. Since there areabout $p(p+1)$ processing cells, the total possible roundoff error is then $p(p+1)|\epsilon x_{max}|$.

In order to prevent the false alarm, a threshold device is needed at the output of e_0 and the threshold has to set at least $|\tau|$. Suppose $\beta = 2, t = 16$, then $u = 2^{-16}$. Given an scaled input data such that $|x_{max}| = 1$, the threshold of a QRD RLS array of order $p = 20$ is

$$
th \ge |\tau|_{max} \simeq 5 \cdot 20 \cdot 21 \cdot 2^{-16} = 0.032. \tag{69}
$$

Since the thresholdis obtained from a conservative derivations, it can always provide a false alarmfree output. However, the estimated threshold may be much higher than that of the actual maximum of the residuals. We can relax the estimated threshold from information obtained in previous data to ensure the threshold will not be too high. A high threshold usually means a small error size may not be able to be detected.

6 Order-Degraded Effects

Consider a order p LS problem with a $n \times p$ complex-valued data matrix $A_p(n)$ denoted by

$$A_p(n) = [\underline{u}(1), \underline{u}(2), \cdots, \underline{u}(n)]^T = [\underline{a}(1), \underline{a}(2), \cdots, \underline{a}(p)] = [A_{p-1}(n) \vdots \underline{a}(p)], \tag{70}$$

a $n \times 1$ desired response vector

$$\underline{y}(n) = [d(1), d(2), \cdots, d(n)]^T,$$

a $p \times 1$ weight vector

$$\underline{w}_p(n) = [w_1^p(n), w_2^p(n), \cdots, w_p^p(n)]^T = [\underline{w}_{p-1|p}(n) \vdots w_p^p(n)]^T, \tag{71}$$

and a $n \times 1$ residual vector

$$\underline{\varepsilon}_p(n) = [e_p(1), e_p(2), \cdots, e_p(n)]^T = A_p(n)\underline{w}_p(n) - \underline{y}(n). \tag{72}$$

Let the index of performance be defined by the weighted l_2 norm of

$$\xi_p(n) = \|\underline{\varepsilon}_p(n)\|_\Lambda^2 = \|\Lambda \underline{\varepsilon}(n)_p\|^2 = \underline{\varepsilon}_p^H(n)\bar{\Lambda}(n)\underline{\varepsilon}_p(n), \tag{73}$$

where

$$\Lambda(n) = diag[\lambda^{(n-1)}, \lambda^{(n-2)}, \cdots, \lambda, 1]$$

with a real-valued forgetting factor $0 < \lambda \le 1$ and $\bar{\Lambda} = \Lambda^2$. Then the least-squares solution, satisfies

$$\xi_{p_{min}}(n) = \min_{\underline{w}} \|\underline{\varepsilon}_p(n)\|_\Lambda^2 = \|A_p(n)\underline{\hat{w}}_p(n) - \underline{y}(n)\|_\Lambda^2. \tag{74}$$

The optimal weight vector can be obtained by solving the normal equation

$$\phi_p(n)\underline{\hat{w}}_p(n) = \psi_p(n), \tag{75}$$

where

$$\phi_p(n) = A_p^H(n)\bar{\Lambda}A_p(n) = \left[\begin{array}{c:c} A_{p-1}^H(n)\bar{\Lambda}A_{p-1}(n) & A_{p-1}^H(n)\bar{\Lambda}\underline{a}_p(n) \\ \hdashline \underline{a}_p^H(n)\bar{\Lambda}A_{p-1}(n) & \underline{a}_p^H(n)\bar{\Lambda}\underline{a}_p(n) \end{array} \right]$$

$$= \left[\begin{array}{c:c} \phi_{p-1}(n) & \underline{\vartheta}(n) \\ \hdashline \underline{\vartheta}^H(n) & \eta(n) \end{array} \right] \tag{76}$$

and

$$\psi_p(n) = A_p^H(n)\bar{\Lambda}\underline{y}(n). \tag{77}$$

It can be easily shown that

$$\xi_{p_{min}}(n) = \|\underline{y}(n)\|_\Lambda^2 - \psi_p^H(n)\hat{\underline{w}}_p(n). \tag{78}$$

For a order $p - 1$ LS problem, then as before, we want to minimize the weighted l_2 norm of

$$\underline{\varepsilon}_{p-1}(n) = [e_{p-1}(1), e_{p-1}(2), \cdots, e_{p-1}(n)]^T = A_{p-1}(n)\underline{w}_{p-1}(n) - \underline{y}(n) \tag{79}$$

with weihgt vector

$$\underline{w}_{p-1}(n) = [w_1^{p-1}(n), w_2^{p-1}(n), \cdots, w_{p-1}^{p-1}(n)]^T.$$

Obviously, the optimal weight vector of order p and $p - 1$ can be obtained as $\hat{\underline{w}}_p(n) = \phi_p^{-1}(n)\psi_p(n)$ and $\hat{\underline{w}}_{p-1}(n) = \phi_{p-1}^{-1}(n)\psi_{p-1}(n)$ respectively. The *difference vector* of the optimal residual vectors of different order is defined by

$$\underline{\Delta}(n) = \hat{\underline{\varepsilon}}_p(n) - \hat{\underline{\varepsilon}}_{p-1}(n) = A_{p-1}(n)[\hat{\underline{w}}_{p-1|p}(n) - \hat{\underline{w}}_{p-1}(n)] + \hat{w}_p^p(n)\underline{a}(p), \tag{80}$$

and the weighted l_2 norm of $\underline{\Delta}(n)$ is defined to be the order degraded performance, $\Gamma(n)$, given by

$$
\begin{aligned}
\Gamma(n) &= \|\underline{\Delta}(n)\|_\Lambda^2 \\
&= \Delta\hat{\underline{w}}^H(n)\phi_{p-1}(n)\Delta\hat{\underline{w}}(n) + \|\hat{w}_p^p(n)\|^2 \cdot \|\underline{a}(p)\|_\Lambda^2 \\
&\quad + 2 \cdot Re[\hat{w}_p^p(n)\Delta\hat{\underline{w}}^H(n)A_{p-1}^H\bar{\Lambda}\underline{a}(p)],
\end{aligned} \tag{81}
$$

where $\Delta\hat{\underline{w}}^H(n) = \hat{\underline{w}}_{p-1|p}(n) - \hat{\underline{w}}_{p-1}(n)$. To relate $\hat{\underline{w}}_{p-1|p}(n)$ with $\hat{\underline{w}}_{p-1}(n)$, we define a LS problem of order $p - 1$ with A_{p-1} as the data matrix and $\underline{a}(p)$ as the desired response. That is, we would like to minimize the weighted l_2 norm of the residual vector

$$\underline{\varepsilon}_{a,p-1}(n) = A_{p-1}(n)\underline{w}_{a,p-1}(n) - \underline{a}(p) = [e_{a,p-1}(1), e_{a,p-1}(2), \cdots, e_{a,p-1}(n)]^T. \tag{82}$$

From (72) and (75), the optimal weight vector can be obtained by solving

$$\phi_{p-1}(n)\hat{\underline{w}}_{a,p-1}(n) = \underline{\theta}(n), \tag{83}$$

where $\underline{\theta}(n)$ is defined in (76). Let the optimal index of performance of this LS problem be $\xi_{a,p-1}(n) = \|\underline{\varepsilon}_{a,p-1}(n)\|_\Lambda^2$, then ϕ_p^{-1} can be represented by

$$
\phi_p^{-1}(n) = \left[
\begin{array}{c|c}
\phi_{p-1}^{-1}(n) & \underline{0}_{p-1} \\
\hline
\underline{0}_{p-1}^T & 0
\end{array}
\right]
+ \frac{1}{\xi_{a,p-1}(n)}
\left[
\begin{array}{c}
-\hat{\underline{w}}_{a,p-1}(n) \\
1
\end{array}
\right]
[-\hat{\underline{w}}_{a,p-1}^H(n) \quad 1]. \tag{84}
$$

Then $\hat{\underline{w}}_p(n)$ can be represented by expressions of order $p - 1$ as

$$
\begin{aligned}
\hat{\underline{w}}_p(n) &= \phi_p^{-1}(n)\psi_p(n) = \phi_p^{-1}(n)
\left[
\begin{array}{c}
A_{p-1}^H(n) \\
\hline
\underline{a}^H(p)
\end{array}
\right]
\bar{\Lambda}\underline{y}(n) \\
&=
\left[
\begin{array}{c}
\phi_{p-1}^{-1}(n)A_{p-1}^H(n) + \frac{\hat{\underline{w}}_{a,p-1}(n)\underline{\varepsilon}_{a,p-1}^H(n)}{\xi_{a,p-1}(n)} \\
\hline
\frac{-\underline{\varepsilon}_{a,p-1}^H(n)}{\xi_{a,p-1}(n)}
\end{array}
\right]
\bar{\Lambda}\underline{y}(n),
\end{aligned} \tag{85}
$$

and therefore,

$$\hat{\underline{w}}_{p-1|p}(n) = \hat{\underline{w}}_{p-1}(n) + \frac{\mathcal{I}(n)}{\xi_{a,p-1}(n)} \hat{\underline{w}}_{a,p-1}(n), \tag{86}$$

$$\hat{w}_p^p(n) = -\frac{\mathcal{I}(n)}{\xi_{a,p-1}(n)}, \tag{87}$$

where $\mathcal{I}(n) = <\hat{\underline{\varepsilon}}_{a,p-1}(n), \underline{y}(n) >_\Lambda$ and $< \underline{x}, \underline{y} >_\Lambda = < \Lambda \underline{x}, \Lambda \underline{y} >$ is the weighted inner product. Thus,

$$\Delta \hat{\underline{w}}(n) = \frac{\mathcal{I}(n)}{\xi_{a,p-1}(n)} \hat{\underline{w}}_{a,p-1}(n). \tag{88}$$

Now, we may proceed to calculate $\Gamma(n)$ in (81). The first term of (81) becomes

$$\Delta \hat{\underline{w}}^H(n) \phi_{p-1}(n) \Delta \hat{\underline{w}}(n)$$

$$= \frac{\|\mathcal{I}(n)\|^2}{\xi_{a,p-1}^2(n)} \hat{\underline{w}}_{a,p-1}(n) A_{p-1}^H(n) \bar{\Lambda} A_{p-1}(n) \hat{\underline{w}}_{a,p-1}(n)$$

$$= \frac{\|\mathcal{I}(n)\|^2}{\xi_{a,p-1}^2(n)} (\hat{\underline{\varepsilon}}_{a,p-1}(n) + \underline{a}(p))^H \bar{\Lambda} (\hat{\underline{\varepsilon}}_{a,p-1}(n) + \underline{a}(p))$$

$$= \frac{\|\mathcal{I}(n)\|^2}{\xi_{a,p-1}^2(n)} (\xi_{a,p-1}(n) + \eta(n) + 2 \cdot Re(< \hat{\underline{\varepsilon}}_{a,p-1}(n), \underline{a}(p) >_\Lambda)). \tag{89}$$

The second term of (81) becomes

$$\|\hat{w}_p^p(n)\|^2 \cdot \|\underline{a}(p)\|_\Lambda^2 = \frac{\|\mathcal{I}(n)\|^2}{\xi_{a,p-1}^2(n)} \cdot \eta(n). \tag{90}$$

The third term of (81) is

$$2 \cdot Re[\hat{w}_p^p(n) \Delta \hat{\underline{w}}^H(n) A_{p-1}^H \bar{\Lambda} \underline{a}(p)]$$

$$= 2 \cdot Re[-\frac{\|\mathcal{I}(n)\|^2}{\xi_{a,p-1}^2(n)} \hat{\underline{w}}_{a,p-1}^H(n) A_{p-1}^H(n) \bar{\Lambda} \underline{a}(p)]$$

$$= -2 \frac{\|\mathcal{I}(n)\|^2}{\xi_{a,p-1}^2(n)} \cdot Re[(\hat{\underline{\varepsilon}}_{a,p-1}(n) + \underline{a}(p))^H \bar{\Lambda} \underline{a}(p)]$$

$$= -2 \frac{\|\mathcal{I}(n)\|^2}{\xi_{a,p-1}^2(n)} \eta(n) - 2 \frac{\|\mathcal{I}(n)\|^2}{\xi_{a,p-1}^2(n)} \cdot Re(< \hat{\underline{\varepsilon}}_{a,p-1}(n), \underline{a}(p) >_\Lambda). \tag{91}$$

Combining (89), (90), and (91) together, we have

$$\Gamma(n) = \frac{\|\mathcal{I}(n)\|^2}{\xi_{a,p-1}(n)} = \frac{\| < \hat{\underline{\varepsilon}}_{a,p-1}(n), \underline{y}(n) >_\Lambda \|^2}{\|\hat{\underline{\varepsilon}}_{a,p-1}(n)\|_\Lambda^2}. \tag{92}$$

Denote the last row of input data matrix $\underline{u}(n) = [\underline{u}_{p-1}(n), u_p(n)]$, the difference of the optimal residual at time n then can be obtained as

$$\|\hat{e}_p(n) - \hat{e}_{p-1}(n)\|^2 = \|\underline{u}^T(n) \hat{\underline{w}}_p(n) - \underline{u}_{p-1}^T(n) \hat{\underline{w}}_{p-1}(n)\|^2$$

$$= \frac{\|\mathcal{I}(n)\|^2}{\xi_{a,p-1}^2(n)} \|\underline{u}_{p-1}^T(n)\hat{\underline{w}}_{a,p-1}(n) - u_p(n)\|^2$$

$$= \frac{\|\mathcal{I}(n)\|^2}{\xi_{a,p-1}^2(n)} \|e_{a,p-1}(n)\|^2. \tag{93}$$

6.1 Geometric Interpretation

From (92), we can see the order degraded performance is indeed the energy of the projection of the desired response $\underline{y}(n)$ onto the subsapce spanned by the optimal residual $\hat{\underline{\varepsilon}}_{a,p-1}(n)$. As the vector $\underline{y}(n)$ becomes more orthogonal to $\underline{\varepsilon}_{a,p-1}(n)$, the order degraded performance is also reduced. Denote the column space of $A_{p-1}(n)$ by $\{A_{p-1}(n)\}$. Then the projection operator $P_{A_{p-1}}$ projects vectors onto space $\{A_{p-1}(n)\}$ and the orthogonal projection operator $P_{A_{p-1}}^\perp = I - P_{A_{p-1}}$ projects vectors onto the space $\{A_{p-1}^\perp(n)\}$ which is orthogonal to space $\{A_{p-1}(n)\}$. The entire space S spanned by $\underline{y}(n)$ can be represented by

$$S = \{A_{p-1}(n)\} \cup \{P_{A_{p-1}}^\perp \underline{a}(p)\} \cup \{A_p^\perp(n)\}, \tag{94}$$

and all of these subspaces are orthogonal to each other. It is obvious that $\hat{\underline{\varepsilon}}_{a,p-1}(n) = P_{A_{p-1}}^\perp \underline{a}(p)$. By projecting the desired response $\underline{y}(n)$ to these subspaces, we obtain

$$\underline{y}(n) = P_{A_{p-1}}\underline{y}(n) + P_{\hat{\underline{\varepsilon}}_{a,p-1}}\underline{y}(n) + P_{A_p}^\perp \underline{y}(n)$$

$$= \underline{y}(n) + < \hat{\underline{\varepsilon}}_{a,p-1}(n), \underline{y}(n) >_\Lambda \frac{\hat{\underline{\varepsilon}}_{a,p-1}(n)}{\|\hat{\underline{\varepsilon}}_{a,p-1}(n)\|_\Lambda^2} + \hat{\underline{\varepsilon}}_p(n), \tag{95}$$

and all of these vectors are also orthogonal to each other. If we drop the vector $\underline{a}(p)$, then the one-dimensional subspace of $\{P_{A_{p-1}}^\perp \underline{a}(p)\}$ cannot be used to represent $\underline{y}(n)$. Therefore, the components of $\underline{y}(n)$ in this subspace is lost and this introduces an error vector $P_{\hat{\underline{\varepsilon}}_{a,p-1}}\underline{y}(n)$ with energy

$$\|P_{\hat{\underline{\varepsilon}}_{a,p-1}}\underline{y}(n)\|^2 = \frac{\|\mathcal{I}(n)\|^2}{\xi_{a,p-1}(n)} = \Gamma(n). \tag{96}$$

The energy of the last component of this error vector is given by (93). Fig.9 illustrates the geometric interpretation discussed above.

7 Summaries

We first introduce the basic concepts of systolic implementation of RLS estimation for adaptive beamformers such as sidelobe canceller and MVDR beamformer. Then, a general algorithm-based fault-toerance is proposed for this class of systolic arrays. Detailed consideration on performance analysis, finite-precision effects, and order-degraded effects are given. In conclusions, the proposed fault-tolerant scheme is robust, and can be designed to be false alarm free without missing error detection problem.

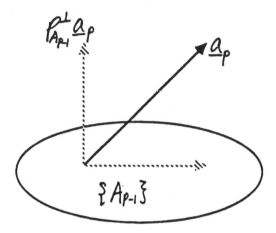

Figure 9: Geometric illustration

References

[1] H. M. Ahmed, J. M. Delosme, and M. Morf, "Highly concurrent computing structures for matrix arithmetic and signal processing", *IEEE Computer*, Vol. 15, No. 1, pp. 65–82, Jan. 1982.

[2] S. T. Alexander, C. T. Pan, and R. J. Plemmons, "Numerical properties of a hyperbolic rotation method for windowed RLS filtering," *Proc. of the IEEE ICASSP*, pp. 423–426, 1987.

[3] Richard Bartels and Linda Kaufman, "Cholesky factor updating techniques for rank 2 matrix modifications," *SIAM J. Martrix Anal. Appl.*, 10(4):557–592, Oct. 1989.

[4] Å. Björck, "Least Squares Methods" in **Handbook of Numerical Analysis Vol. II:** Finite difference methods - Solution of equations in R^n. Elsevier North Holland, 1989.

[5] A. W. Bojanczyk and F. T. Luk, "A novel MVDR beamforming algorithm," *SPIE Vol. 26 Advanced Algorithms and Architecture for Signal Processing II*, pp. 12–16, 1987.

[6] R. T. Compton Jr., **Adaptive antennas: Concepts and performance,** Prentice Hall, 1988.

[7] W. Morven Gentleman, "Least squares computations by Givens transformations without square roots," *J. Inst. Maths Applics*, 12:329–336, 1973.

[8] W. M. Gentleman and H. T. Kung, "Matrix triangularization by systolic array," *Proc. SPIE*, Vol. 298: Real-time signal processing IV, pp. 19–26, 1981.

[9] W. Murray P. E. Gill, G. H. Golub and M. A. Saunders, "Methods for modifying matrix factorizations," *Mathematics of Computation*, 28(126):505–535, Apr. 1974.

[10] G. H. Golub and C. F. Van Loan, **Matrix computations,** Johns Hopkins Press, 2nd ed., Baltimore, MD, 1989.

[11] S. Haykin, **Adaptive filter theory,** Prentice-Hall, Englewood Cliffs, NJ, 1986.

[12] D. E. Heller and I. C. F. Ipsen, "Systolic networks for orthogonal decompositions," *SIAM J. Sci. Stat. Comput.*, No. 4, pp. 261–269, 1983.

[13] S. F. Hsieh and K. Yao, "Hyperbolic Gram-Schmidt pseudo orthogonalization with applications to sliding window RLS filtering," *Proceedings of 24th Annual Conference on Information Sciences and Systems*, pp. 197–202, Mar. 1990.

[14] S. F. Hsieh and K. Yao, "Systolic implementation of windowed recursive LS estimation," *Proc. of IEEE Int'l Symp. on CAS*, pp. 1931–1934, May 1990.

[15] S. Kalson and K. Yao, "Systolic array processing for order and time recursive generalized least-squares estimation," *Proc. SPIE 564, Real-Time Signal Processing VIII*, pp. 28–38, 1985.

[16] H. T. Kung, "Why systolic architectures?" *IEEE Computer*, Jan. 1982.

[17] S. Y. Kung, et al., *Proceedings of the Int'l Conf. on Application Specific Array Processors*, IEEE Press, Sept. 1990.

[18] C. L. Lawson and R. J. Hanson, **Solving least squares problems,** Prentice-Hall, Englewood Cliffs, N.J., 1974.

[19] F. Ling, D. Manolakis, and J. G. Proakis, "A recursive modified Gram-Schmidt algorithm for least-squares estimation," *IEEE Trans. on Acous., Speech, and Signal Processing*, Vol. ASSP-34, No. 4, pp. 829–836, Aug. 1986,

[20] K. J. R. Liu, S. F. Hsieh, and K. Yao, "Two-level pipelined implementation of systolic block Householder transformations with application to RLS algorithm," *Proc. Int'l Conf. on Application-Specific Array Processors*, pp.758–769, Princeton, Sep. 1990.

[21] J. G. McWhirter, "Recursive least-squares minimisation using a systolic array," *Proc. SPIE 431, Real time signal processing VI*, pp. 105–112, 1983.

[22] J. G. McWhirter and T. J. Shepherd, "Systolic array processor for MVDR beamforming," *IEE Proceedings*, Vol. 136, Pt. F, No. 2, pp. 75–80, Apr. 1989.

[23] R. A. Monzingo and T. W. Miller, **Introduction to adaptive arrays,** John Wiley & Sons, Inc., 1980.

[24] C. M. Rader and A. O. Steinhardt, "Hyperbolic Householder transformations," *IEEE Trans. Acoust., Speech, Signal Processing*, Vol. ASSP-34, No. 6, pp. 1589–1602, Dec. 1986.

[25] R. Schreiber, "Implementation of adaptive array algorithms," *IEEE Trans. on Acoust., Speech, Signal Processing*, Vol. ASSP-34, No. 5, pp. 1038–1045, Oct. 1986.

[26] B. D. Van Veen and K. M. Buckley, "Beamforming: a versatile approach to spatial filtering," *IEEE ASSP Mag.*, Vol. 5, No. 2, pp. 4–24, Apr. 1988.

[27] B. Widrow and S. D. Stearns, **Adaptive signal processing,** Prentice-Hall, Englewood Cliffs, NJ, 1985.

[28] J. H. Wilkinson, **The algebraic eigenvalue problem,** Oxford University Press, England, 1965.

[29] B. Yang and J. F. Böhme, "On a systolic implementation and the numerical properties of a multiple constrained adaptive beamformer," *Proc. of the IEEE ICASSP*, pp. 2819–2822, 1989.

[30] C.R. Ward, P.J. Hargrave and J.G. McWhirter, "A novel algorithm and architecture for adaptive digital beamforming", IEEE Trans. Antennas Propagat., Vol AP-34, pp.338-346, March, 1986.

[31] C.-Y. Chen and J.A. Abraham, "Fault-tolerant systems for the computation of eigenvalues and singular values", Proc. SPIE, Vol 696, Advanced Algorithms and Architectures for Signal Processing, pp.228-237, 1986.

[32] C.-Y. Chen and J.A. Abraham, "Current error detection in VLSI processor arrays", Proc. SPIE, Vol 826, Advanced Algorithms and Architectures for Signal Processing II, pp. 205-214, 1987.

[33] K.-H. Huang and J.A. Abraham, "Algorithm-based fault-tolerance for matrix operations", IEEE Trans. Computer, Vol C-33, pp.518-528, June, 1984.

[34] J.-Y Jou and J.A. Abraham, "Fault-tolerant matrix arithmetic and signal processing on highly concurrent computing structures", Proc. IEEE, Vol 74, pp.732-741, May, 1986.

[35] S.Y. Kung, **VLSI Array Processors,** Prentice Hall, 1988.

[36] F.T. Luk and H. Park, "Fault-tolerant matrix triangulization on systolic array", IEEE Trans. Computer, Vol 37, pp.1434-1438, Nov. 1988.

[37] S.-W. Chan and C.-L. Wey, "The design of concurrent error diagnosable systolic arrays for band matrix multiplications", IEEE Trans. CAD, Vol 7, pp.21-37, Jan. 1988.

[38] J.A. Abraham et al., "Fault tolerance techniques for systolic array", IEEE Computer, Vol 20, pp.65-76, July 1987.

[39] J.V. McCanny and J.G. McWhitter, "Some systolic array developments in the United Kingdom", IEEE Computer, Vol 20, pp.51-64, July 1987.

[40] K.G. Shin and T.-H. Lin, "Modeling and measurement of error propagation in a multimodule computing system", IEEE Trans. Computer, Vol 37, pp.1053-1066, Sep. 1988.

[41] C.J. Anfinson and F.T. Luk, "A linear algebraic model of algorithm-based fault tolerance", Proc. IEEE Int'l Conf. Systolic Array, pp.483-493, May 1988.

[42] A. Avizienis, "Fault tolerant computing - An overview", IEEE Computer, Vol 4, pp.5, Jan. 1971.

[43] J.A.B. Fortes and C.S. Raghavendra, "Gracefully degradable processor arrays", IEEE Trans. Computer, Vol C-34, pp.1033-1044, Nov. 1985.

[44] S.Y. Kung et al., **VLSI and Mordern Signal Processing**, Prentice-Hall, 1985.

[45] H.T. Kung and M.S. Lam, "Wafer-scale integration and two-level pipelined implementaiton of systolic array", J. Parallel Distrib. Comput., Vol 1, pp.32-63, 1984.

[46] H. Lev-Ari and B. Friedlander, "On the systematic design of fault-tolerant processor arrays with application to digital filter", Proc. IEEE Workshop on VLSI Signal Processing, pp.483-494, Nov. 1988.

[47] S.N. Jean, C.W. Chang and S.Y. Kung, "Graceful degradation schemes for static/dynamic wavefront array", Proc. Int'l Conf. Parallel Processing, pp.249-255, Aug. 1988.

[48] J.-Y. Jou and J.A. Abraham, "Fault-tolerant algorithms and architectures for real time signal processing", Proc. Int'l Conf. Parallel Processing, pp.359-362, Aug. 1988.

[49] J.-Y. Jou and J.A. Abraham, "Fault-tolerant matrix operation on multiple processor system using weigted checksums", Proc. SPIE Vol 495 Real Time Signal Processing VII, pp.94-101, 1984.

[50] S.Y. Kung, C.W. Chang, and C.W. Jen, "Real-time reconfiguration for fault-tolerant VLSI array processors", Proc. Real-Time Systems Symposium, pp.46-54, 1986.

[51] I. Koren and D.K. Pradham, "Yield and performance enhancement through redundancy in VLSI and WSI multiprocessor systems", IEEE Proc. Vol 74, pp.699-711, May, 1986.

[52] I. Koren and M.A. Breuer, "On area and yield considerations for fault-tolerant VLSI processor arrays", IEEE Trans. Computer, Vol C-33, pp.21-27, Jan. 1984.

[53] C.J. Anfinson et al., "Algorithm-based fault-tolerant techniques for MVDR beamforming", Proc. IEEE ICASSP, pp.2417-2420, May, 1989.

[54] K.J.R. Liu and K. Yao, "Gracefully degradable real-time algorithm-based fault-tolerant method for QR recursive least-squares systolic array", Proc. International Conference on Systolic Array, pp. 401-410, Killarney, Ireland, May, 1989.

[55] K.J.R. Liu, "Dynamic range for finite-precision QRD LS algorithm and its stability", Proc. IEEE Int'l Sym. Circuits and Systems (ISCAS), pp.3142-3145, New Orleans, May 1990.

[56] M. Sami and R. Stefanelli, "Reconfigurable architectures for VLSI processing arrays", Proc. IEEE, pp.712-722, May 1986.

[57] M. Chean and J.A.B. Fortes, "A texonomy of reconfiguration techniques for fault-tolerant processor arrays", IEEE Computer, Vol. 23, pp.55-69, Jan. 1990.

5

Parallel Computation of Fan Beam Back-Projection Reconstruction Algorithm in Computed Tomography

Wen-Tai Lin, Chung-Yih Ho, and Chi-Yuan Chin

Corporate Research and Development
General Electric Company
P.O. Box 8, KWC 417
Schenectady, NY 12301

Abstract

This paper describes parallel processing algorithms and architecture for computed tomography (CT) using a fan beam energy source and, more particularly, a processor array adapted for computing the back-projection reconstruction algorithm (BPR). By making use of some inherent parallelism of BPR, a modular ring architecture in multiple of eight processors is shown to be well suited for a class of CT algorithms. The parallel architecture has encompassed a reconfigurable binary tree to accommodate a wide range of processor interconnections needed for different image processing schemes, ranging from front-end filtering, calibration, and FFT to post-processings such as artifact removal and image display.

Section 1 Introduction

When the first commercial CT machine was constructed, the algorithm it used was called algebraic reconstruction technique (ART). Basically, ART solves a system of linear equations iteratively. Because of the intensive computation involved, this method was soon replaced by a direct Fourier technique (DFT) based on the Central Slice Theorem. It then evolved into the convolution and back-projection reconstruction (BPR), which is the most widely used method in commercial CT scanners nowadays. The fan beam energy source, which simplified the scanning geometry and largely reduced the scanning time, was soon adopted in the CT machine. Although the DFT method has an inherent speed advantage over back-projection reconstruction, it is considered to be unsuitable for use with the fan beam scanner because of excessive sensitivity to noise [1]. The BPR is more suitable for view pipelining and yields images that are relatively free of artifacts from processing.

In general, the required processing speed is determined by the speed of rotation gantry and the frequency of diagnostic request. As the need for more sophisticated three-dimensional (3D) image reconstruction increases, the conventional approach, which uses a minicomputer as a host and attaches optional hardware to speed up the BPR processes, is experiencing difficulties in keeping up with the large amount of data generated from multiple scans; the approach requires $O(N^4)$ operations. The bottleneck appears to be in the reconstruction unit, where BPR is carried out in a sequential (or pipelined) manner. Although rebinning (followed by DFT) has been considered to be the most promising technique for parallelizing the image reconstruction of a fan beam CT scanner, its image quality is still a problem because of the need for interpolating data on a Cartesian coordinate from data over a polar coordinate. The rebinning process itself requires $O(KN^3)$ operations for compensating truncation in the interpolation of rebinned projection data, where the size of K also determines the quality of the reconstructed image [2].

In this paper we present a method for parallelizing the image reconstruction of a fan beam CT scanner. The result of this method is also applicable to 3D cone beam CT machines if the BPR method is also used for image reconstruction on each plane. A reconfigurable architecture is proposed to nail down the issues of cost-effectiveness, scalability, modularity, fault tolerance, and performance. In Section 2 we briefly introduce the BPR method and review the computation steps involved in some commercial CT reconstruction units. In Section 3 an inherent parallelism associated with BPR is explored. By making use of such parallelism, a modular ring architecture in multiple of eight processors is proposed. Then in Section 4, we present a generic architecture using binary trees to integrate multiple rings and to provide reconfigurability and fault tolerance for the system. Section 5 presents the conclusions of this effort.

Section 2 Fan Beam CT Computation

Figure 1 is a graph of the geometry of fan beam scanning and back-projection. Assume that a complete scan is comprised of N views evenly taken from a full rotation of the gantry and out of each view there are M samples collected from the detector array. Then after certain pre-convolution and calibration the input to the BPR process can be organized as a $N \times M$ data array D. Let $f(r, \theta)$ be density of an image pixel located at a polar coordinate (r, θ). The reconstruction of $f(r, \theta)$ can be summarized as follows:

$$f(r, \theta) = \sum_{n=1}^{N} W_n D_{n,k} \tag{1}$$

$$W_n = \frac{\beta_{n+1} - \beta_{n-1}}{2U^2} \tag{2}$$

$$U^2 = [r \cos(\beta_n - \theta)]^2 + [E + r \sin(\beta_n - \theta)]^2 \tag{3}$$

$$k = \alpha^{-1} \tan^{-1} \{r \cos(\beta_n - \theta)/[E + r \sin(\beta_n - \theta)]\} \tag{4}$$

The equations are drawn from [3, 4]. For each pixel, Equation (1) accumulates N–weighted detector values collected from N projection angles. Equation (2) and (3) calculate W_n, the nth weight and U^2, the square distance of pixel (r, θ) and the x-ray energy source radiating at angle β_n. Equation (4) estimates a relative position (in the detector space) which lines up the x-ray source and the pixel of interest. If k is not an integer, then a more accurate value of $D_{n,k}$ may be obtained by interpolation using adjacent detector values.

The CT reconstruction can be divided into three major computation steps (see Figure 2): pre-convolution, convolution and back-projection, and post-processing. The computation time in each major step is estimated based on a set of algorithms including several pre-filtering and calibration schemes (T_1), FFT (T_2), BPR (T_3), and an image restoration algorithm to remove artifacts introduced by inaccurate interpolation (T_4). For a moderate image resolution (512 x 512), T_3 accounts for 80% of total computation time based

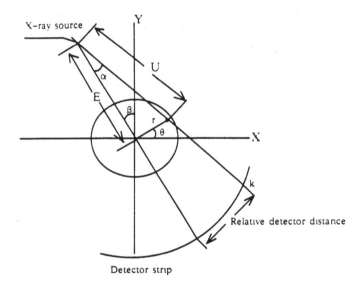

Figure 1 The geometry of fan beam scanning and back-projection.

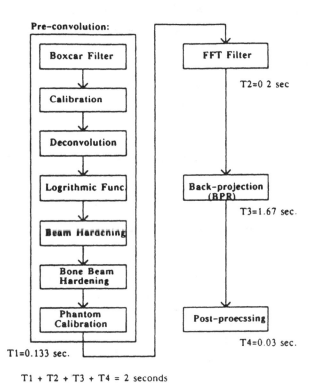

Figure 2 The computation steps of a CT fan beam reconstruction unit.

on a sequential machine. This figure is obtained by assuming that Equations (2) to (4) are computed elsewhere.

Figure 3 further illustrates the dataflow of each computation step. A downward (left-to-right) arrowhead signals that the computation is carried out along each individual column (row). There are only two types of computation flows (column- and row-directions) involved in the pre-convolution and FFT stages. At the back-projection step, the data in the view-detector (V-D) space are transformed into a rectangular image space. In a raster scan display, the computation flow can be viewed as random input from the V-D space and sequential output to an image space. In the most extreme case, the post-processing may also be treated as random access in the image space. The general rule-of-thumb is to provide a flexible architecture to accommodate a wide range of image processing.

Section 3 Parallelizing the Fan Beam Computations

The time complexity for reconstructing an L-by-L-pixel image from an N-view-by-M-detector projection data is $O(L^2 N)$. To improve the system performance at relatively low cost, one promising approach is to explore the parallelism of BPR. Given P identical processors (PEs), one would like to assign almost equal amount of task to each processor with minimum resource redundancy. The first step is to reduce the trigonometric computations by using the property of trigonometric symmetry.

Let k_1 and W_1 be the coefficients evaluated for the reconstruction of pixel $f(r_1, \theta_1)$ using view data obtained from projection angle β_1. Then k_1 and W_1 may be reproduced for reconstructing many different pixels lying on the same concentric locus. This concept could be derived from Equation (3) and (4) by showing that if $\sin(\beta_i - \theta_i) = \sin(\beta_1 - \theta_1)$ and $\cos(\beta_i - \theta_i) = \pm \cos(\beta_1 - \theta_1)$, then $W_i = W_1$ and $k_i = \pm k_1$. The relationship between $\beta_i - \theta_i$ and $\beta_1 - \theta_1$ can be expressed as follows:

$$\beta_i - \theta_i = (\beta_1 - \theta_1), \quad \text{or} \tag{5}$$
$$\beta_i - \theta_i = \pi - (\beta_1 - \theta_1). \tag{6}$$

Now the $N \times M$ projection data array is partitioned into P consecutive segments in view direction, followed by assigning each segment to a processor's local memory, e.g. PE_1 holding row 1 through row N/P, and so on. Then

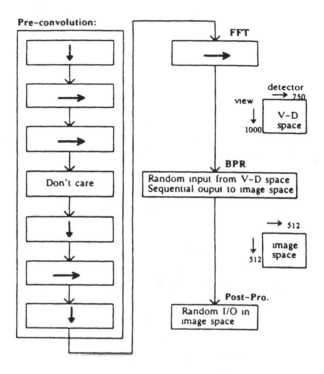

Figure 3 The directions of dataflow in a view-detector space.

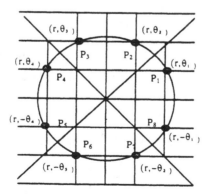

Figure 4 Eight concentric grid points in a rectangular coordinate.

Equation (1) can be split into P partial sums as follows:

$$f(r_1, \theta) = \sum_{n=1}^{N/P} W_n D_{n,k} + \cdots + \sum_{n=\frac{N}{P}(P-1)+1}^{N} W_n D_{n,k}. \qquad (7)$$

By keeping all these segmented projection data in local memory, PE_i, for $i = 1, ..., P$, will always be responsible for computing the ith partial sum of some pixel $f(r_1, \theta_j)$. Moreover, as long as the Equation (5) holds, only one coefficient pair, e.g. (k_1, W_1), has to be computed. Naturally, a ring-connected processor array would serve the need for accumulating and transferring all these simultaneously computed partial sums.

For image display using raster scan, only the pixels located at the grid points of a rectangular coordinate need to be reconstructed. The eight concentric grid points (P_1 through P_8) shown in Figure 4 can be split into two sets: the even-indexed points and the odd-indexed points. From Equation (5) one sees that if the odd-indexed points (P_1, P_3, P_5 and P_7) use a coefficient pair (k, W), then the even-indexed points will use $(-k, W)$. Consequently, there are always four or eight grid points lying on the same concentric locus and the image is naturally partitioned into eight octants. This allocation of image data is particularly useful when ring artifacts are to be removed right after the back-projection stage.

To speed up the back-projection process further, one could introduce more PEs, in an increment of eight, to the system. For example, 32 PEs can be organized as an 8-row-by-4-column processor array. The projection data are partitioned into 32 consecutive segments and assigned to the processors in a row-major sequence. Now four boundary units are required to compute four pairs of (k, W) coefficients. With each column processors sharing the same (k, W) coefficients, up to 32 pixels can be simultaneously reconstructed. And, again, the output image is naturally partitioned into 32 suboctant regions (see Figure 5).

Section 4 A Generic Fan Beam CT Architecture

Based on the current BPR package, a fan beam CT reconstruction machine has to compute at a rate no less than 2 billion operations (i.e., multiplications and additions) per second in order to boost the resolution to 1000-by-1000 pixels per image at a reconstruction rate of one cross section per second. This estimation has not even taken the computation

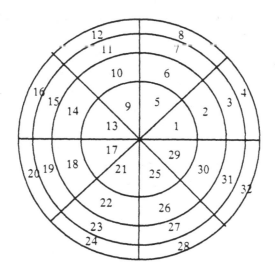

Figure 5 Partitioning the entire image into 32 sub-octant regions.

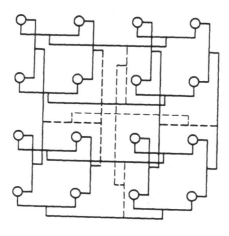

Figure 6 A 16-node modified orthogonal tree.

of (k, W) pairs into account. To derive a generic parallel architecture for fan beam CT, one has to consider a large category of signal processing algorithms that are being used in the current CT reconstruction package. The key issues are (1) processor interconnection, (2) distributed memory, (3) system control, and (4) system software. Based on the pre-processing steps described in Figure 2, it is more favorable to partition the projection data along the view dimension; this not only avoids the data transfer when computing 1D FFT (which is being used as a convolution filter along the detector dimension), but also makes the search of detector values confined in each local memory during the BPR stage. However, there are still substantial amount of computation involved in interchanging data between processors, such as convolution in view direction, BPR (which basically needs a ring), and post-processings. Intuitively, one sees that an architecture based on the mesh-connected topology would probably meet most of the needs. However, for reasons soon to become clear, we introduce a modified orthogonal trees (MOT) as a superset of the mesh- and ring-connected architectures.

A conventional orthogonal tree (also called mesh-of-trees) is constructed by organizing a processor array into columns of trees and rows of trees. It was originally proposed as a general-purpose parallel computer ideal for VLSI implementation [5], which has only recently been received as a vision-processing architecture [6]. An MOT is formed by combining these subtrees into one column tree and one row tree (see Figure 6). Let P be a power of 2. In [5] it is shown that when P processors are connected to the leaf cells of a tree, they can be viewed as being interconnected into a ring and are able to rotate data in $O(\log P)$ time steps. The key advantage of this configuration is that it may bypass any length of masked processors in $O(\log P)$ time. Likewise, when an $P^{1/2} \times P^{1/2}$ processor array is interconnected by an MOT, it becomes very reliable and dynamic. Below are some useful computation structures based on a P-node MOT network:

1. Mesh-connected arrays are formed when P processors are connected to the leaf cells of the column and row trees according to some predetermined column and row sequences.

2. Given a tree with P leaf cells, up to $2^{(P/2-1)}$ of different node sequences in the same ring can be formed. Therefore when two orthogonal trees are overlaid as a P-node MOT network, the number of possible interconnection patterns is $2^{(P-2)}$.

3. Fault-tolerant arrays of fixed structures (such as mesh array) can, in general, be handled as follows:

a. reject the faulty node by simply disconnecting it from the MOT;
b. explore a new set of fault-free nodes that can preserve the interconnection topology;
c. if spatial reconfiguration is impossible, then use time redundancy to implement the remaining connections at a second phase.

In the literature there is a great deal of fault-tolerance-related research based on the mesh-connected array. Because of the scope of this paper, we shall only focus on the construction of MOT and on the principles of partitioning image-processing tasks based on this type of architecture.

A Binary Tree Implemented with Shuffle Busses

Because of the $O(\log P)$ time steps involved in moving data from node to node, it is very desirable to speed up the data transfer rate in an MOT network. In [7] we designed a shuffle bus that can be viewed as a bidirectional, word-parallel, pipeline bus using a programmable swapping mechanism to achieve data redistribution. A binary tree-structured shuffle network (BTS) is configured as having each of the parent nodes connected to a left child (Lc) and a right child (Rc) via a multiplexer and a shuffle node. To operate as a ring bus, the links of its BTS are separated into two groups (Lc and Rc) and alternately activated. In Figure 7 we show that two sets of data, say {a,b,c,d} and {e,f,g,h}, can be rotated in a pipelined fashion. The tree has one additional shuffle node preceding the root. Immediately after the first set of data are loaded, all the left edges are activated (designated as L-operation) and rotations are done in each of these shuffle bus segments. It is then followed by an R-operation, where all of the right edges are activated and another rotation is carried out in each of these segments. At the beginning of the third cycle, a second set of data, {e,f,g,h}, are loaded to the leaf nodes. Although four macro-steps are needed to finish the rotation of each data set, its throughput rate is doubled by keeping two data sets on the same BTS. To reverse the rotation direction, the R-operation is executed first, then followed by the L-operation; it is repeated in this manner until the desired number of shifts is achieved. Note that, for a P-node ring bus to be implemented on a BTS, the longest linear shuffle segment is $\log P + 2$. Hence the time it takes to do one rotation is $4(\log P + 1)T_s$, where T_s is the swapping time between adjacent nodes. Note that typical T_s on shuffle bus is around 5 ns.

To further speed up the ring interconnections on a BTS, one could form a wraparound BTS, as shown in Figure 8, where each leaf node is connected

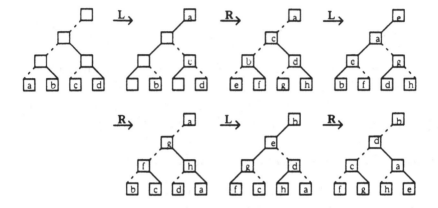

Figure 7 Pipelining two data sets on a binary shuffle tree.

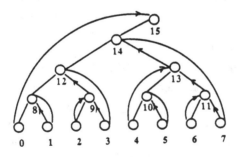

Figure 8 A wraparound binary tree.

to a corresponding intermediate node such that the original linear segments are now formed as small rings. In so doing, each R- or L-operation is finished in one clock. This speedup reduces the ring bus cycle time from $4(\log N + 1)T_s$ to $4T_s$.

The selection of MOT as an interconnection network for CT parallel processors has been based on two criteria: (1) it should be reconfigurable to include a wide range of CT post-processing algorithms (besides the ones shown in Figure 2) and (2) the operation of data transfer over the network should be simple and easy to maintain. As mentioned before, a P-node MOT can be configured up to 2^{P-2} different interconnection patterns, which is done by programming the control codes applied to the multiplexers that are used to direct the intermediate tree nodes. To ease the system control and programming issues, the network is monitored by an independent control agent, which implements, from task to task, a configuration table either prepared during compile time or dynamically scheduled at run time. The processors only talk to the network through local FIFOs. For nonsystolic type of data transfers, the data elements (or packets) can be tagged with destination codes. Then by traversing (or recirculating) data through the rings, the system can be programmed as a dataflow machine.

Partitioning Image Processing Tasks

Because of a large volume of input (or intermediate) data involved, it is more favorable to obtain parallelism by partitioning the input data and allowing each processor to work on its own subimage. For example, a smooth systolic 2D convolution-like algorithm requires that the convolution size be equal to the number of processors. By partitioning the entire image into subimages and storing each subimage in the local memory of a separate PE, one can derive a semisystolic dataflow that is not hampered by heavy cross-boundary data transfers. In general, the image processing can be categorized as follows:

1. For algorithms that are homogeneously applied (by sliding an operation kernel) across the entire image, each processor is assigned to work on its own subimage. When it comes to crossing a subimage boundary, the kernel is split into 2 (for 1D image partition) or 4 (for 2D image partition) wraparound subkernels within the subimage area of each PE. Hence, rather than shuffling input data across the processor boundaries, intermediate results are forwarded to their destination processors. The

network traffic problem is solved by evenly distributing the same amount of traffic over the entire computation period.

2. For heterogeneous operations, PEs are normally assigned with different tasks; intermediate results are combined in a dynamic fashion; and the processors are dynamically grouped into separate rings to carry out independent tasks. A dynamically reconfigurable network is essential for this type of image processing. On the other hand, preserving the locality of these image processors is also highly desirable, since such property helps to confine the local traffic in the local region of a network. In the most extreme case, all of the PEs are grouped to form a pipeline architecture and the memory banks are coordinated by one memory controller. The advantage of this approach is to avoid irregular partitioning of original data such as the case of rebinning in particular or image warping in general.

Examples of the first category are CT back-projection reconstruction, convolution, and pattern matching; examples of the second category are FFT, histogram computations, image warping, and interpolation.

Section 5 Conclusions

In this paper we show that the conventional fan beam CT algorithms can be parallelized by sharing the (k, W) pairs with eight processors. We also proposed an 8-node ring as a fundamental processor module to make use of this inherent concurrence. To include a wide variety of image post-processing algorithms, we took a further step by upgrading the one-dimensional ring to a two-dimensional MOT network. By establishing quick bypassing paths through the trees, the processors can be dynamically grouped, thereby facilitating global communications. On the other hand, since a tree can be decomposed into subtrees, the locality of smaller rings is also well preserved.

Bibliography

[1] J.E. Wheeler, et al. "Instant Computed Tomography Reconstruction". In Internal Document of General Electric Company, April 1984.

[2] Hui Peng. "Fan Beam Reconstruction in Computer Tomography from Full and Partial Projection Data". PhD Dissertation: Rensselaer Polytechnic Institute, New York, 1988.

[3] G.T. Herman, et al. "Convolution Reconstruction Techniques for Divergent Beams". Computer of Biologic Medicine, 6(10), October 1976.

[4] B.K. Gilbert, et al. "Rapid Execution of Fan Beam Image Reconstruction Algorithms Using Efficient Computational Techniques and Special Processors". IEEE Tran. Biomedical Engineering, BME-28(2), Feb. 1981.

[5] J.D. Ullman. "Computational aspect of VLSI". Computer Science Press, 1984.

[6] Quentin F. Stout. "Mapping Vision Algorithms to Parallel Architectures". In Proceedings of the IEEE, volume 76, August 1988.

[7] W.T. Lin and J.P. Hwang. "A High-Speed Shuffle Bus for VLSI Arrays". IEEE Journal of Solid-State Circuits, February 1988. Also in the Proceedings of the 1987 Symposium on VLSI Circuits, Karuizawa, Japan, May 1987.

6

Affine Permutations of Matrices on Mesh-Connected Arrays

Björn Lisper
Department of Telecommunication and Computer Systems
P.O. Box 70043
S-100 44 Stockholm, SWEDEN
bjornl@tds.kth.se

Sanjay Rajopadhye
Computer Science Department
University of Oregon
Eugene, OR 97403-1202, USA
sanjay@cs.uoregon.edu

Abstract

We present methods to permute matrices in mesh-connected uniform square arrays with local control only. The permutations that we consider form a class called *affine permutations*, which includes transpose and many other row/column reorderings. We first present a general scheme where destination tags are generated on the fly, and a standard sorting algorithm on the mesh is used to route the elements to their respective destination. We then specialize to four operations: in-place reflection, in-place rotation, and on-the-fly versions of these (that permute the matrix while it is loaded into the array), and show how they can be implemented very efficiently with local control only. We also develop a general theoretical model for prescheduled data transfers in distributed systems. This model can be applied to permutations, and we use it to verify one of the specialized operations.

1 Introduction

Parallel computing invariably includes the transmission and rearrangement of large quantities of data. Solutions to the communication problem range from global but not very scalable solutions, like buses, through somewhat more scalable solutions like interconnection networks [Sie85], to local but scalable point-to-point connection schemes. There is a corresponding range in the type of parallellism exploited, from large-grain computation on a few, powerful processors to fine-grain computation on many simple units. The development of

VLSI technology has stimulated research at the latter end of the spectrum. The following properties are desirable for on-chip integration of parallel systems:

- *Regular layout.* If the system consists of many similar units (or *cells*), the layout for the cell can be replicated. Thus, uniformity is desirable.

- *Local communication.* If the system has nearest-neighbour connections only, then long, area-consuming wires with large capacitance are avoided. Thus, the absence of such wires will make the circuit more compact and also enable it to run faster.

- *Local control.* Local communication also implies local control, possibly with the exception of a global clock. This restricts the possible actions of cells, since control information must reach a cell before it can change the cell's behaviour.

- *1D and 2D topologies.* VLSI chips are essentially planar surfaces. Only one and two dimensional topologies can be laid out on planar surfaces without violating the nearest neighbour restriction. Regular 1D topologies are linear arrays and rings. Regular 2D topologies are mesh connected arrays, possibly with diagonal connections.

- *I/O restricted to boundaries.* Locality of communication implies that I/O can take place only at the boundary of the array. This is a constraint both on data and control.

Together, these properties define the concepts of synchronous *systolic arrays* [KL80] and asynchronous *wavefront array processors* [Kun88]. Systolic (wavefront) array implementations have been suggested for many algorithms, especially in areas like linear algebra and image and signal processing. See for instance [MMJ89,MMU87]. There has also been a great interest in formal methods for synthesizing systolic implementations of regular algorithms, especially by space-time mapping methods [CS84], [Che86], [DI87], [HL87], [JRK87], [RFS85], [Lis89], [MW84], [Mol82], [MF86], [Qui84], [RK88], [Raj89].

With the exception of [DI87], where systems of coupled recurrence equations are considered, the work in the field of synthesis has concentrated on implementations of single algorithms. The question immediately arises, however, how to interface different systolic algorithms. Such interfacing could take place in space, i.e. connecting several systolic arrays into one system, or in time, i.e. combining different operations taking place in the same array at different times, or a combination of both.

If two systolic arrays are linked, some transformation of the output of the first must usually be done to fit the input pattern of the second. n output elements can be transformed by interconnection networks with $O(n \log n)$ binary switches [Sie85] to provide the proper inputs to the next array, but these networks are not amenable to VLSI implementation. *Cellular permutation networks* with up to n^2 switches [OO87] are better suited for this, but they are costly in space and the capability to perform general permutations is often not needed for interfacing systolic computations. It is also wasteful to have one input for each element. For systolic array operations, $O(n)$ inputs are usually required to arrive in a pipelined fashion of $O(\sqrt{n})$ streams with $O(\sqrt{n})$ elements each. Thus, a pipelined network with $O(\sqrt{n})$ inputs and outputs is better suited to interface systolic arrays. Some pipelined interfacing operations, like transposition of streams, were considered by O'Leary [O'L87]. Optimization of a class of pipelined buffers for the interfacing of systolic arrays has been described by Wah, Aboelaze and Shang [WAS88].

Interfacing of systolic computations taking place in the same array has not, to our knowledge, been described, but related work has been done for data permutations in mesh-connected SIMD computers [Fla82,NS80]. In SIMD systems, however, the conditions are not the same as for systolic/wavefront systems. In SIMD machines the program flow of each processor is centralized, which is not in accordance with the systolic demand on local control. On the other hand, all the processors in a SIMD system must perform the same operation at a given time, which does not necessarily hold in a systolic/wavefront type system where a new instruction, or a control signal, may flow through and cause the processors on its way to change behaviour.

In this paper, we will consider *affine* permutations of matrices on a mesh-connected square systolic or wavefront array. Such permutations arise often in practice: examples are matrix transpose and many row and column re-orderings. All the operations described here can be implemented in a systolic fashion with local control only. We will consider two types of permutations: in-place permutations, where data resides in the array before and after the operation, and "flow" permutations, where the rearrangement of data is carried out "on-the-fly" while being loaded into the array over a boundary. By a simple transformation, the latter class of operations can be used for on-the-fly permutations while *unloading* the data as well. This type of conversion provides an alternative to the interfacing of arrays with separate buffers.

The remainder of this paper is organized as follows. In Sec. 3 we shall describe a general technique that enables us to perform arbitrary affine permutations of square matrices. We then describe a few drawbacks of this approach, and identify four transformations that are important from a practical viewpoint. Then, in Sec. 4, we develop a theoretical framework for reasoning about ensembles of data storage elements that can route data locally amongst themselves. The theory is general enough to describe any prescheduled data transfers. Sec. 5 then describes an architecture for reflecting a matrix in-place, and illustrates how the theory can be used to verify the correctness of the implementation. This architecture is further refined in Sec. 6 so that the implementation uses only local control; an array for on-the-fly reflection is also presented. The next two sections present arrays for rotation. Finally, we present our conclusions.

2 Preliminaries

In this paper we will use *binary relations* and operations on them. A binary relation R over a set A is a subset of $A \times A$. If $\langle a, b \rangle \in R$ we write aRb, and we call a an *immediate predecessor* of b. Every binary relation R over A can be seen as a directed graph $\langle A, R \rangle$, where there is an edge from a to b exactly when aRb. The *transitive and reflexive closure* of R is denoted R^*: it holds that aR^*b exactly when there is a path of length ≥ 0 in $\langle A, R \rangle$ from a to b. For any relation R on A and elements a, b in A we define $n(R, a, b) = \{\, c \mid aR^*cR^*b \,\}$, i.e., the set of intermediate nodes in all paths from a to b.

Z_n denotes the *ring* of integers modulo n, $Z_n = \{0, 1, \ldots, n-1\}$, with binary operators $+_n$ (addition modulo n) and $\cdot$ (multiplication modulo n, also denoted by $(ab)_n$). Since Z_n is a ring, multiplication does not have an inverse, although many elements are invertible (precisely those that are relatively prime to n, there are as many as $O(n)$ of them). $Z_n^{k \times k}$ denotes the $k \times k$ matrices whose elements are in Z_n. We can define matrix addition and multiplication as ususal—all scalar additions and multiplications are modulo n. The determinant $|M|$ of a matrix in $Z_n^{k \times k}$ can also be defined. It is easy to show that $Z_n^{k \times k}$ is also a ring, in which an element M has an inverse iff $|M|$ has an inverse in Z_n, and that there are $O(n^{k^2})$ such elements.

In designing systolic arrays, one typically starts with an initial algorithm that specifies the problem. This algorithm is normally expressed as the computation of a function at all points in an *index-space* (*viz* the integer lattice points in a subset of Euclidean space). A mathematical description of such an algorithm is given by the following definition.

Definition 1 *A* Recurrence Equation *over a domain* $\mathcal{D}_n$ *is defined to be an equation of the form*

$$f(p) = g(f(q_1), f(q_2) \ldots f(q_k))$$

> *where* $p \in \mathcal{D}_n$; $\mathcal{D}_n$ *is a convex polyhedral subset of* Z^m
>
> n *is a size parameter*
>
> $q_i \in \mathcal{D}_n$ *for* $i = 1 \ldots k$;
>
> *and* g *is a single-valued function, strictly dependent on all its arguments.*

A system of recurrence equations is a set of l mutually recursive such equations, defining functions $f_1, f_2, \ldots f_l$.

A Recurrence Equation *of the form defined above is called a* Uniform Recurrence Equation *(URE) if* $q_i = p + b_i$, *for* $i = 1, \ldots, k$, *where the* b_i*'s are constant m-dimensional integer vectors. It is said to be an* Affine Recurrence Equation *(ARE) if for* $i = 1, \ldots, k$, $q_i = A_i p + b_i$, *where the* A_i*'s are constant $m \times m$ integer matrices, and b_i's are constant $m \times 1$ integer vectors.*

A modular affine recurrence equation *(MARE) is an ARE where* $\mathcal{D}_n = Z_n^m$, *and for* $i = 1, \ldots, k$, $A_i \in Z_n^{m \times m}$, *and* $b_i \in Z_n^{k \times 1}$. *A MARE where* $b_i = 0$ *is called a* modular linear recurrence equation *(MLRE). Note that since a total* *order cannot naturally be defined over rings, it does not make sense to talk about inequalities, and hence about convex polyhedral domains. The domain of computatin for MAREs and MLREs is thus taken to be the entire index space, Z_n^k (which is finite).*

3 An Architecture for arbitrary modular affine permutations

The problem of implementing systems of affine recurrences (AREs) on regular, systolic arrays has received considerable attention. It has been shown [RF90,RK88] that systolic arrays are characterized by systems of recurrences that have *uniform* rather than affine dependencies, (such *uniform* recurrence equations (UREs) form a proper subset of AREs—where the linear part of the dependency is the identity). There has therefore been an effort towards automatically converting AREs into equivalent UREs. The case when the affine map is rank deficient has now received a satisfactory answer (see [Raj89], [QV89], [WD89], [RTRK88] for example). There has also been work addressing the case when the dependency is non-singular. Culik and Fris [CF85] describe a technique called "folding of the plane" which has been extended by Yaacoby and Cappello [YC88] to a "generalized fold," which permits the localization of such dependencies. Furthermore, as suggested by Roychowdhury et al. [RTRK88], a non-singular dependency can be converted into a singular one by extending the dimensions of the index space, although this approach (often unnecessarily) extends the domain of computation.

While it is an interesting problem to investigate how the localization methods mentioned above can be extended to MAREs and MLREs, in this paper we shall only consider a particular subclass of MAREs, namely those with a single, invertible dependency, A. The recurrences that we study have the form

$$f(p) = f'(Ap + b)$$

Such MAREs capture the essence of non-singular modular dependencies, and the architecture that we propose here can be easily adapted to arbitrary MAREs where there is more than one dependency and the computation is not merely the identity function. Moreover, unlike the previous work, we are interested in deriving a *single* architecture that can implement this entire class of MAREs. Because of practical considerations, we consider the domain to be two-dimensional, although our architecture can be easily extended to higher dimensional domains. Thus, our problem may be specified as follows: Given a square $n \times n$ matrix, M compute the $n \times n$ matrix N given by

$$N_{i,j} = M_{A[i,j]^T + b} \tag{1}$$

As mentioned earlier, we are interested in designing two variants of the final architecture — one for the case when the input matrix is being fed into the array, and one for the case when the matrix has already been loaded into the array. Moreover, we are interested in achieving this *without* the use of any global control (for example, the decision to start a certain computation is made by a processor solely through some control signals which may only be propagated locally).

The problem that we are addressing is a proper subset of the *permutation routing problem*: given a processor array where each processor has a data packet and a destination address (based on a 1-1 permutation of the processor labels), determine an algorithm to route all the packets to their correct destinations. Our problem is a proper subset in the sense that since there are only $U(n^4)$ distinct non-singular dependency matrices, and only $O(n^2)$ distinct vectors, b, the total number of permutations that we consider is only $O(n^6)$, which is a very small fraction of the $n^2!$ possible permutations. It is well known that the routing problem can be directly solved by sorting on the destination addresses. Algorithms for achieving this in $O(n)$ steps have been presented by Kung and Thompson [TK77], and also by Nassimi and Sahni [NS79]. Recently Leighton et al. [LMT] have given a time-optimal algorithm that uses $2n - 2$ steps and constant local memory. In most of this work however, it is assumed that the destination addresses are initally present in the processors. This is a valid assumption if the processor array is being used to simulate a PRAM (a shared memory random access parallel machine), but is not very practical in our context. Indeed, if the permutations are determined off-line, there may be considerable effort involved in simply informing each processor of the destination of it packet. The architecture that we present is essentially an array that achieves this "address calculation" task in $2n$ time steps. By a simple modification, this can be reduced to n. After this initial stage, the Leighton et al. routing algorithm is used to achieve the desired permutation.

Since our permutation is an affine permutation, the final destination of the matrix element at processor $[i,j]^T$ is $A[i,j] + b$. Note that this is an affine function of the processor co-ordinates, $[i,j]$. Thus, if we design a mesh connected processor array where each processor computes an affine function of its space co-ordinates, the result of this computation is precisely the *address tag* that we need to sort on. Hence we will first design a square systolic array where each processor can compute an affine function of its co-ordinates. We must do this without having the processors to use any global information, and without explicitly "informing" the processors of their locations (otherwise the processors cannot be considered to be identical). Our method is based on the following observation, which follows directly from distributivity of our ring.

Remark 1 *Given,* $x = A[i,j]^T + b$, *the value of an affine map at a point,* $[i,j]^T$, *for any constant vector* ρ, *the value of the affine map at* $[i,j] + \rho$ *is*

$$x + A\rho$$

Since ρ is a *constant* vector, so also is $A\rho$, and this indicates that at all points in the domain, $A[i,j] + b$ can be computed *incrementally*. The idea is similar to the data pipelining techniques proposed by Rajopadhye [Raj89], but

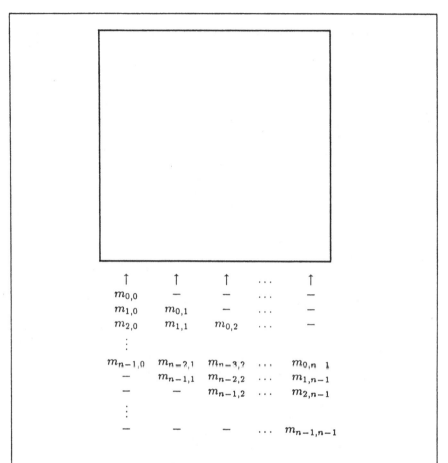

Figure 1: Using data pipelining to compute the destinations of the matrix elements

here we are interested in computing the dependency itself. As in the case of data pipelining, we must choose vectors ρ which form a basis for the space that we want to span, i.e., index space $[i,j]$. In our case we simply choose the two vectors $[0,1]$ and $[1,0]$. We also observe that $A[0,0] + b$ is b, so we can simply use the translation part of the affine map to "initialize" the pipeline.

As a result, we obtain the architecture shown in Fig 3, which consists of a square array on $n \times n$ processors and an extra row below it. All processors (in the $n \times n$ square) have an accumulator in which they can store the two coordinates of an index vector, and an output register (going vertically up) which can store similar index values. They also have arithmetic hardware that can perform addition of such co-ordinate values (modulo n). They all have an input bus connected to the output register of the processor immediately below, and behave as follows. They start computing when they recieve a *start* control signal from the processor immediately below them (this signal is propagated upwards at a speed of one processor per clock cycle). When they get the start signal, they load their accumulator and output register from the input bus. From then on, at each clock tick, they add the value on the input bus to their accumulators and also latch it on the output register. They stop when they receive another control signal (from their left neighbor, propagated to the right at a speed of one processor per clock cycle). The stop signals reach all the processors in a column at the same time, and are synchronized to reach a column exactly one time unit after the topmost processor has received the start signal. Thus, by the time that the stop signal arrives, the topmost (zero-th) processor has performed no additions on its accumulator, the next one (first) has performed one, and the i-th one, i. Given such operation of the processors, it is clear that if the very first input (i.e., the input that arrives with the start signal) to the i-th processor in the j-th column is $A[0, j] + b$ and all subsequent inputs are at $A[1, 0]$, this will also be true for the processor immediately above it, and hence, inductively for the entire column (provided we can show that it holds for the base case — the $(n-1)$-th row). Thus, by the time the stop signal arrives, the i-th processor will have added $A[1, 0]$ i times to $A[0, j]$ and hence will have computed, $A[0, j] + b + i * A[1, 0]$, i.e., $A[i, j] + b$, which is precisely what is required.

The bottom row of processors is used to ensure that the base case mentioned above holds. These processors are responsible for computing $A[0, j] + b$ for their respective columns. Each processor has an accumulator and an output register, and has two inputs (from the accumulator and output register of its *left* neighbor, respectively). The accumulator also serves as input to the processor immediately above it. The processor begins computing when it receives a control signal from its left neighbor (the control signal is sent to the right as well as upwards). When it receives the start signal, the processor loads its accumulator with the sum of its two inputs and its output register with the value of the second input. From then on, it simply loads its accumulator from the first input. Thus we want the following invariants to hold:

- At the *end* of the first operative clock cycle (i.e, the cycle in which the start signal arrives) of any processor in this row, its accumulator contains $A[0, j] + b$ and its output register contains $A[0, 1]$.

- At all subsequent times, its accumulator contains $A[1, 0]$.

Proving this inductively is straightforward, given the operation of the processors above. The base case will be ensured if the first and second inputs to the leftmost processor at the first time instant are $b - A[0, 1]$ and $A[0, 1]$ respectively (their sum will then be b which is exacltly what is required), and the first input is $A[1, 0]$ from then on. Alternatively, the first set of inputs could be b and $A[0, 1]$ and the processor be specialized so that it performs *no* addition at the first time step.

In the above architecture, we have used an extra row of processors, but this was done merely to aid the exposition. Since the computation performed by the bottom two rows is mutually exclusive (as is the hardware needed for it), we very easily merge their functionality into one processor. This one again has two registers, *acc*, and *out*, and can read the two registers of its left neighbor. Moreover, the input bus of the processor above it is connected to both, *acc* and *out* thhrough a multiplexor. When it gets the control signal, it loads *acc* with the sum of its two inputs, the *out* register with its second input, and sets the mux so that the processor above will read the contents of *acc* on the next cycle. From the next cycle on, it adds the second input to *acc* and also loads it into *out* and sets the mux so that *out* is sent to the processor above it.

We thus have an architecture which enables a square mesh-connected array to compute a modular affine map of the processor space. Such an array could be used with only a simple modification, for either of the two variants — for the case when the data is not already present in the array, we will simply need to provide an additional register in each processor for the actual data values (such a register is needed to store the data value, in any case). When the start signal arrives, the processor begins to simply propagate the incoming data, and when the stop signal arrives, it merely latches the input data into this register.

The next phase is to actually route the data to their correct destinations by sorting the tags computed in the first part*. To do this we use the Leighton et al. routing algorithm. It may seem at first glance that becuase the destination values are computed incrementally over the array, it is possible to overlap the routing phase with the destination computation. However, this is misleading, and does not improve the performance. In the worst case, the very last address that is computed (at PE $[n-1, n-1]$) may turn out to be $[0, 0]$, so $2n - 2$ additional time units are needed and this is how much the routing algorithm would have taken if everyone started routing at this instant.

The time required for the address calculation phase is $2n$ steps. However, this does not fully utilize the available boundaries. Indeed, the maximum distance of any point from a boundary is $n/2$, and if we partition the problem into four independent problems, one for each quadrant, and have the processors propagate the results in the appropriate directions, the time can eb reduced to n steps.

3.1 Specialized Architectures

The architecture presented above is general in the sense that it can perform *arbitrary* affine permutations of square matrices. However, there is a price to be paid for this generality. The time complexity of the arrays is larger – the time required to compute the destination addresses is $2n$, and the additional time to sort is $2n - 2$. Moreover, the array is not easily extensible, since most of the computation performed by the processors is dependent on the problem size (all the addition is modulo n, and the sorting algorithm requires that the processors be aware of their co-ordinates and of n, etc.). We therefore investi-

---

*Thus the signal that we have called the stop signal so far, merely indicates the start of this second phase, and a more accurate name would be *sort* signal.

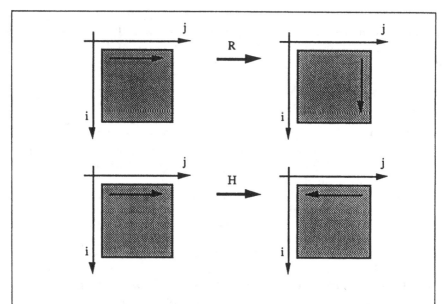

Figure 2: Illustration of 90 degree right rotation and horizontal reflection.

gate in the remainder of this paper, a small class of specialized permutations. In particular, we consider permutations consisting of reflections and 90 degree rotations. We anticipate that such transformations will have practical applications in fast manipulation of frame buffers and in image processing. We first note that because the space of such permutations forms a group, all the transformations can all be performed simply by composing the *generators* of the group. Specifically, the two basic permutations that we shall use as our generators are (note that all arithmetic operations are modulo n):

- 90 degree right rotation (assuming a coordinate system with j increasing to the right and i increasing downwards, as shown in Fig 2 below.

$$R(i,j) \equiv \begin{pmatrix} 1 & 0 \\ 0 & -1 \end{pmatrix} \begin{pmatrix} i \\ j \end{pmatrix} \qquad (2)$$

- Horizontal reflection (about the middle of the array):

$$C(i,j) \equiv \begin{pmatrix} 0 & 1 \\ -1 & 0 \end{pmatrix} \begin{pmatrix} i \\ j \end{pmatrix} \qquad (3)$$

As an example, matrix transposition $(i,j) \mapsto (j,i)$, specified by the affine map

$$A = \begin{pmatrix} 0 & 1 \\ 1 & 0 \end{pmatrix}$$

can be implemented as a right rotation followed by a horizontal reflection. It can also be seen that

$$\begin{pmatrix} 0 & 1 \\ 1 & 0 \end{pmatrix} = \begin{pmatrix} 1 & 0 \\ 0 & -1 \end{pmatrix} * \begin{pmatrix} 0 & 1 \\ -1 & 0 \end{pmatrix}$$

For both of these permutations, we are interested in designing hardware structures that can perform them when the data is already in the array (called in-place arrays) and when the data is initially off-line (flow-arrays). In the next section we outline a theoretical framework for reasoning about prescheduled data transfers. The four arrays are described in the four subsequent sections.

4   Prescheduled permutations of data

Efficient specialized architectures for a certain permutation can be found if the data transfers are prescheduled. When the data transfer pattern in decided upon in advance, the architecture can be tailored to perform that particular pattern efficiently. Usually, much of the control overhead can then be avoided. In this section we will hint at a theoretical model to describe permutations of data through prescheduled data transfers. The details can be found elsewhere [LR]. We will use this model to verify some of the presented specialized permutation architectures.

In general, we are interested in permutations of indexed data fields $a(i)$, where i belongs to some finite index set I. The permutations are then permutations of the index set. We view data as being stored in memory and permutations taking place there. Thus, we assume an *address space M*. An address in M could possibly consist of several fields: in a distributed system, one field may give the processor location and another field a local memory address.

We do not, however, want to restrict the theory to permutations from addresses to addresses. In a systolic array, for instance, data fields may flow over boundaries and permutations may have to take place between flows. Since a flow of data indicates different data appearing at the same place at different times, this means that the temporal aspect also is important. Here we restrict our attention to synchronous systems: an event in such a system can be identified with an address and a discrete time. If we take the times to be the nonnegative integers $\{0, 1, \ldots\} = N$, then the set of all events is the *address space-time $N \times M$*. Instead of associating a datum statically with an adress, we can identify it with events in address space-time which enables us to describe systems where data fields move. Especially, we want to specify "interfaces" in address space-time where each item in a data field has a unique space-time coordinate. Such interfaces specify where and when data is input, and where and when data is to be picked up. This leads to the following definition:

Definition 2 An injective function from an index set to an address space-time is a *space-time mapping* of the index set.

If i is mapped to $\phi(i) = \langle t, m \rangle$ by a space-time mapping ϕ, the interpretation is that a data item $a(i)$ indexed by i is to occur at memory address m at time t. Next, we need a way to specify how data is propagated in (a possibly distributed) memory with time.

Definition 3 A relation $\rightarrow$ on the address space-time $N \times M$ is a *transfer relation* if it fulfils the following:

1. For all $\langle t, m \rangle$, $\langle t', m' \rangle$ in $N \times M$, $\langle t, m \rangle \rightarrow \langle t', m' \rangle \Rightarrow t < t'$. (causality)

2. $\rightarrow$ is a forest. (uniqueness of source)

The transfer relation $\rightarrow$ *connects* s to s' iff $s \rightarrow^* s'$. Transfer relations are used to model prescheduled movements of data. Intuitively, if $\langle t, m \rangle \rightarrow \langle t', m' \rangle$, then the data item stored in address m at time t becomes stored in m' at time t'. The causality property in Def. 3 rules out the transfer of data backwards in time. Forests are disjoint unions of trees: thus, elements in forests have unique immediate predecessors. Property 2 then ensures that two different data items never are written to the same address at the same time. When an address space-time event is connected to another event there is a transfer relation path between them, which means that data will be transferred from the first to the second in a number of steps.

Definition 4 Let π be a permutation of the index set I. Let ϕ and ψ be space-time mappings from I to an address space-time. Let $\rightarrow$ be a transfer relation on the address space-time. Then $\rightarrow$ *implements* π with respect to ϕ and ψ if the following holds:

1. For all $i \in I$, $\phi(i) \rightarrow^* \psi(\pi(i))$ (Connectivity).

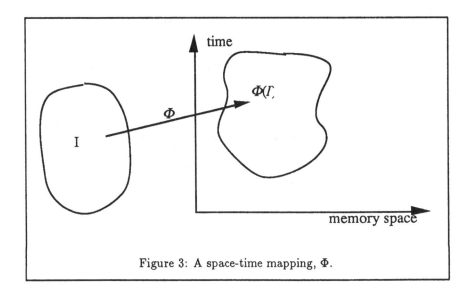

Figure 3: A space-time mapping, Φ.

2. For any $i, j \in I$, $i \neq j$, $n(\rightarrow, \phi(i), \psi(\pi(i))) \cap n(\rightarrow, \phi(j), \psi(\pi(j))) = \emptyset$.
   (Disjoint paths)

Property 1 says that there is a (unique) path in $\rightarrow$ from $\phi(i)$ to $\psi(\pi(i))$, i.e. the value, say $a(i)$, at $\phi(i)$ will be copied to $\psi(\pi(i))$. Property 2 states that the paths between different input and outputs never overlap. Thus, a value that is input will never overwrite a value being transferred to an output through the transfer relation. Def. 4 can be visualized by a "commuting diagram" (see Fig. 5. We refer to ϕ and ψ as the input and output space-time maps, respectively, and say that ϕ, ψ and $\rightarrow$ constitute a "space-time permutation".

Property 2 allows the possibility that $s \rightarrow \phi(i)$ for some s. This may seem somewhat awkward, since a strictly operational interpretation of $\rightarrow$ would imply the value from s being transferred to $\phi(i)$ with a resulting write conflict. The problem is, however, only in the interpretation. The correct interpretation is that the data input at $\phi(i)$ overwrites whatever should have been transferred to there. We have chosen the definition here because of its mathematical simplicity.

The model of transfer relations also can be used to reason about other types of data transfers than permutations. To show that a value is broadcast, for instance, amounts to finding a path from the event where it is input to the

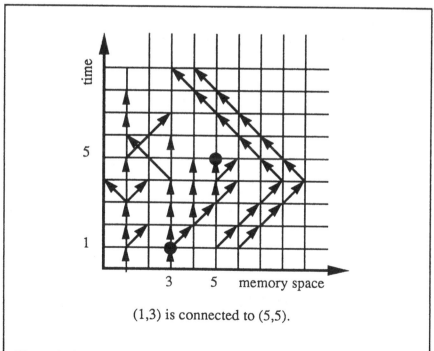

(1,3) is connected to (5,5).

Figure 4: A transfer relation connecting two points in address space-time.

respective events where it is to be present.

An interesting task is to *verify* that a given transfer relation really implements a permutation w.r.t. two space-time mappings. Typically, the transfer relation will then be derived from some more or less formal description of the architecture implementing the permutation. Such a verification is carried out in section 5, where the transfer relation is specified on a case by case basis, through a set of linear inequalities informally derived from the supposed action of an array of processing elements. The verification essentially amounts to showing that there is a transfer path in address space-time from any input to its output.

Transfer relations can describe both a *desired* action of an architecture, i.e. constitute a specification, and describe the *actual* action of a given architecture. An example of this is found in section 5. This opens up the interesting possibility of *synthesis*, through refinement of "specification" transfer relations into "implementation" transfer relations. More formally, we say that a transfer relation $\rightarrow_1$ is *implemented* by the transfer relation $\rightarrow_2$ iff any path in $\rightarrow_1$ is a path in $\rightarrow_2$.

A transfer relation that describes an implementation must be reasonably close to the implementation. First, in a synchronous memory system the contents of a memory cell is copied and re-stored every clock cycle. Thus, if $\langle t, m \rangle \rightarrow \langle t', m' \rangle$ it must hold that $t' = t + 1$ when $\rightarrow$ describes such a system. We call this property *temporal locality*. Second, a memory cell could possibly not be connected to all other memory cells. Especially, we are interested in systems where there is a spatial locality and a memory cell is connected to nearest neighbours only. Spatial locality constraints often arise in distributed systems, where an address is divided into a processor address and a local ad-

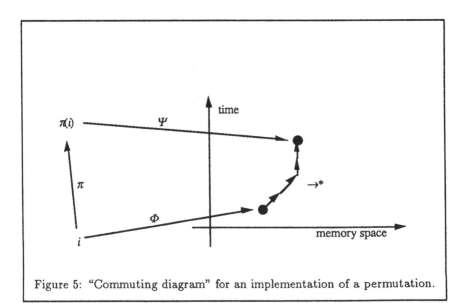

Figure 5: "Commuting diagram" for an implementation of a permutation.

dress. We call an address space *processor-decomposable* if it can be written as a cartesian product $P \times L$. If the relation $\leftrightarrow$ on P describes spatial locality amongst processors, then we call the transfer relation $\rightarrow$ *spatially local* w.r.t. $\leftrightarrow$ iff $\langle t, (p, l) \rangle \rightarrow \langle t', (p', l') \rangle$ always implies $p \leftrightarrow p'$. We finally say that $\rightarrow$ is *implementable* w.r.t. $\leftrightarrow$ iff it is temporally local and spatially local w.r.t. $\leftrightarrow$.

A locality constraint can also be imposed on the *control* of a memory system. We want to model systems where the only global control signal is the clock, and the behaviour of a processor (or cell) is completely controlled by locally propagated signals. Certain cells are *boundary* cells; they can recieve external control signals and pass them on. Denote the set of boundary cells by B, and the *distance* from B to the cell p, i.e. the shortest $\leftrightarrow$-path from any cell in B to p, by $d(B, p)$. Then p can clearly not be affected by an external control signal, arriving at time t, before $t + d(B, p)$. This gives a restriction on the transfer relations that such a system can implement, since a cell in such systems must recieve a control signal to change its behaviour. The issues of verification and synthesis are treated more in detail in [LR].

In the following sections, we use the concepts defined here to describe a number of permutation operations of a two-dimensional $n \times n$ mesh-connected processor array, and to verify one of the operations. In such an array, an address consists of a local address and a two-dimensional processor address (x, y): $0 \leq x, y \leq n - 1$. Every local address corresponds to a local register. Two processors (cells) (x, y), (x', y') will be adjacent (i.e. $(x, y) \leftrightarrow (x', y')$) whenever $|x - x'| + |y - y'| \leq 1$. The distance between two cells (x, y), (x', y') is simply $|x - x'| + |y - y'|$. (x, y) is a boundary cell if $x = 0$, $x = n - 1$, $y = 0$ or $y = n - 1$. A cell (x, y) then has the distance x, $n - 1 - x$, y and $n - 1 - y$ to the respective boundaries, with corresponding restrictions on the control.

5 Architecture H1: Horizontal Reflection of a Stored Matrix

In the data transfer model developed in the previous section, the in-place, horizontal reflection is specified by the following transfer relation:

$$\langle 0, i, j, \text{Acc} \rangle \rightarrow \langle T, i, n - j - 1, \text{Acc} \rangle \tag{4}$$

where the input and output maps are respectively given by $\phi(i, j) = \langle 0, i, j, \text{Acc} \rangle$ and $\psi(i, j) = \langle T, i, j, \text{Acc} \rangle$ (for some, as yet unspecified time, $T > 0$). It is easy to see that this is a transfer relation: causality is obvious, because $T > 0$; and since $[i, n - j - 1]$ is a permutation of $[i, j]$, $\rightarrow$ is a forest with each arc being a distinct tree.

Our implementation will be a refinement of this relation which satisfies the properties of spatial and temporal locality. We first observe, that all the forests in the relation may be partitioned into distinct planes that are parallel to the i axis. Hence, all transfers could be restricted to the same *row* of memory cells, and therefore, we will design a *one-dimensional* array of cells which can independently perform the horizontal reflection of one row of the matrix. We

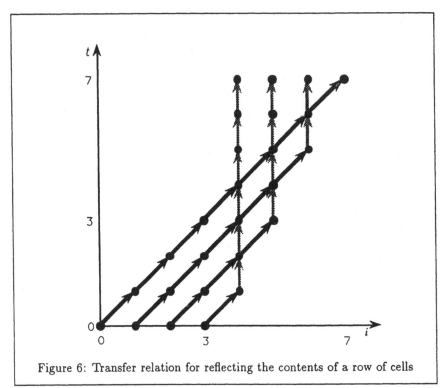

Figure 6: Transfer relation for reflecting the contents of a row of cells

first informally describe the operation of the array, then express its behavior as a transfer relation, and then prove that the transfer relation is a correct implementation of Eqn 4. Each processor in the linear array has three storage elements, labelled *Acc*, *DL* and *DR* (denoting Accumulator, Data-Left and Data-Right, respectively). A cell, y, can read DR of its left neighbor, $y - 1$, and DL of its right neighbor, $y + 1$. At $t = 1$, all cells in the *left* half of the row ($2y < n$) load their DR from Acc of $y - 1$ (the leftmost cell does an undefined, dont-care load), and the *right* half cells load their DL from Acc of $y + 1$. From then on, the left half cells copy DR of cell $y - 1$ to their own DR, until $t = y$, do nothing (undefined, dont-care) until $t = n - 2y - 1$, at which time they copy DL of cell $y + 1$ into their Acc, and then until $t = n - y - 1$ they copy DL of $y + 1$ to their DL. From $t = n - 2y$ to $t = n$ they must also ensure that their accumulator remains unchanged. The right half cells are similar, but move data to the *left*. This yields the transfer relation shown in Fig 6, and it is easy to see that the data values Acc of cell y at $t = n - 1$ is the value that was in Acc of $n - y - 1$ at $t = 0$. The transfer relation is formally defined below (Eqns 5 - 14).

$$(t = 1 \wedge 2y < n) \quad \Rightarrow \quad \langle t-1, y-1, \text{Acc} \rangle \rightarrow \langle t, y, \text{DR} \rangle \tag{5}$$

$$(t > 1 \wedge t \leq y \tag{6}$$

$$\wedge\, t > 2y - n - 2) \quad \Rightarrow \quad \langle t-1, y-1, \text{DR} \rangle \rightarrow \langle t, y, \text{DR} \rangle \tag{7}$$

$$(t = 1 \wedge 2y \geq n) \quad \Rightarrow \quad \langle t-1, y+1, \text{Acc} \rangle \rightarrow \langle t, y, \text{DL} \rangle \tag{8}$$

$$(t > 1 \wedge t \leq n - y - 1 \tag{9}$$

$$\wedge\, t > n - 1 - 2y) \quad \Rightarrow \quad \langle t-1, y+1, \text{DL} \rangle \rightarrow \langle t, y, \text{DL} \rangle \tag{10}$$

$$t = n - 2y - 1 \quad \Rightarrow \quad \langle t-1, y+1, \text{DL} \rangle \rightarrow \langle t, y, \text{Acc} \rangle \tag{11}$$

$$t = 2y - n + 1 \quad \Rightarrow \quad \langle t-1, y-1, \text{DR} \rangle \rightarrow \langle t, y, \text{Acc} \rangle \tag{12}$$

$$(t > 2y - n - 2 \,\wedge \tag{13}$$

$$t > n - 1 - 2y \wedge t < n) \quad \Rightarrow \quad \langle t-1, y, \text{Acc} \rangle \rightarrow \langle t, y, \text{Acc} \rangle \tag{14}$$

To show that the above definition constitutes a transfer relation, we must show that it is causal and forms a forest. Causality is obvious from the definition, since the RHS of each of the equations above are of the form $\langle t-1, m \rangle \rightarrow \langle t, m' \rangle$. Moreover, the above definition is an implementation, since each of the instances are temporally and spatially local (spatial locality also follows from the fact that the relations are of the form $\langle t, y, L \rangle \rightarrow \langle t', y \pm 1, L' \rangle$). Hence $\rightarrow$ is a transfer relation if we can show that it is a forest, i.e., there do not exist two distinct points in space-time, $\langle t', y', L' \rangle$ and $\langle t'', y'', L'' \rangle$, which are both related to the same point $\langle t, y, L \rangle$. To do this, we need to simply ensure that for all different equations defining a variable (i.e., equations whose RHS is of the form $\langle t, y, \text{Var} \rangle$), the guards are disjoint. This can be verified by straightforward inspection of the *regions* defined by the (in)equalities: for example, the intersection of $(t = 1 \wedge 2y < n)$ and $(t > 1 \wedge t > 2y - n - 2 \wedge t \leq y)$ can be seen to be null.

We will now prove that Eqns 5 - 14 are an implementation of Eqn 4, with respect to the input and output maps $\phi(j) = \langle 0, j, \text{Acc} \rangle$ and $\psi(j) = \langle n-1, j, \text{Acc} \rangle$. We need to show that $\langle 0, j, \text{Acc} \rangle \rightarrow^{*} \langle n-1, n-1-j, \text{Acc} \rangle$. The RHS of this consists of points on the $t = n - 1$ line, and since it involves Acc, this can only arise from one of Eqns 11, 12 or 14. Of these, the guard for Eqn 11 is true for $t = n - 1$ only at $y = 0$, and that of Eqn 12 can be true only at $y = n - 1$. At all other points, (i.e., $\langle t, y \rangle$ such that $0 < y < n - 1, t = n - 1$), the value of Acc must be as defined by Eqn 14. Now, within the region defined by the guard of this equation, we may draw a family of straight lines parallel to $\langle -1, 0 \rangle$ (corresponding to $\langle t, y \rangle - \langle t-1, y \rangle$) and these lines correspond to transitive closure of $\rightarrow$ within the region. Because of causality, these lines *must* intersect another boundary of the region, and by some algebraic manipulation, we see that the line from $\langle n-1, j \rangle$ meets $\langle n - 2j - 1, j \rangle$ (if $2j < n$), or $\langle 2j - n - 1, j \rangle$ (if $2j \geq n$). Moreover, the two additional points, $\langle n - 1, 0 \rangle$ and $\langle n - 1, n - 1 \rangle$ also belong to these two line segments, respectively. Hence we have shown that

$$2j < n \quad \Rightarrow \quad \langle n - 2j - 1, j, \mathrm{Acc} \rangle \rightarrow^* \langle n - 1, j, \mathrm{Acc} \rangle \tag{15}$$

$$2j \geq n \quad \Rightarrow \quad \langle 2j - n + 1, j, \mathrm{Acc} \rangle \rightarrow^* \langle n - 1, j, \mathrm{Acc} \rangle \tag{16}$$

$$\tag{17}$$

Since the points $\langle n - 2j - 1, j \rangle$ and $\langle 2j - n + 1, j \rangle$ are precisely the regions defined by the guards of Eqns 11 and 12, we may easily conclude that

$$2j < n \quad \Rightarrow \quad \langle n - 2j - 2, j + 1, \mathrm{DL} \rangle \rightarrow^* \langle n - 1, j, \mathrm{Acc} \rangle \tag{18}$$

$$2j \geq n \quad \Rightarrow \quad \langle 2j - n, j - 1, \mathrm{DR} \rangle \rightarrow^* \langle n - 1, j, \mathrm{Acc} \rangle \tag{19}$$

$$\tag{20}$$

Proceeding in this manner, and reasoning about the domains of Eqns 7 and 10 (using lines of slope $\langle -1, -1 \rangle$ and $\langle -1, 1 \rangle$ respectively) we can show that

$$2j < n \quad \Rightarrow \quad \langle 1, n - j - 2, \mathrm{DL} \rangle \rightarrow^* \langle n - 1, j, \mathrm{Acc} \rangle \tag{21}$$

$$2j \geq n \quad \Rightarrow \quad \langle 1, n - j, \mathrm{DR} \rangle \rightarrow^* \langle n - 1, j, \mathrm{Acc} \rangle \tag{22}$$

$$\tag{23}$$

Once again, the regions defined by the guards of these equations make the guards of Eqn 8 and 5 true, and we therefore have

$$2j < n \quad \Rightarrow \quad \langle 0, n - j - 1, \mathrm{Acc} \rangle \rightarrow^* \langle n - 1, j, \mathrm{Acc} \rangle \tag{24}$$

$$2j \geq n \quad \Rightarrow \quad \langle 1, n - j - 1, \mathrm{Acc} \rangle \rightarrow^* \langle n - 1, j, \mathrm{Acc} \rangle \tag{25}$$

$$\tag{26}$$

Since the union of the implicants of both these equations is a tautology, and the implied formulas are identical, this reduces to

$$\langle 0, n - j - 1, \mathrm{Acc} \rangle \rightarrow^* \langle n - 1, j, \mathrm{Acc} \rangle \tag{27}$$

which is precisely what is required.

Although the proof presented above was not very formal, we make some remarks regarding the proof. Note that Eqns 5 - 14 have a very distinctive form. They can, in fact, be shown to be equivalent to UREs. Hence, we can make use of the convexity properties of the domain (remember that all the guard expressions are conjunctions of linear (in)equalities) and reduce the verification problem to reasoning about families of lines in such polyhedra. We anticipate that this will permit us to automate much of this process.

It is also apparent that the architecture presented above is not practical. For example, it requires each cell to receive *broadcast* signals indicating start and stop instants, and should also know (or be informed through another broadcast signal) which half they are on. Since we desire that this should be done without

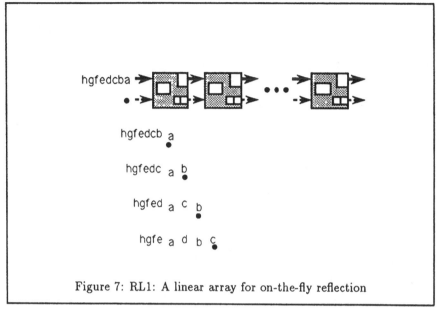

Figure 7: RL1: A linear array for on-the-fly reflection

any global signals, and all I/O must occur at the boundaries, we can only do this "row by row." Hence, the time required to complete the operation will be $2n$. In the following section we will present an implementation that fits this characterisation.

6 Practical Architectures for Reflection

We shall now develop practical architectures for reflecting a matrix (both in-place and on-the-fly versions). Rather than describing the development in a completely formal manner, we rely on a somewhat informal presentation. The interested reader can easily check that the verification technique used in Sec 5 can be used to verify the correctness of our designs. We first note that reflection of a matrix can be decomposed into *independent* reflections of its individual rows (because the *row index* of any matrix element is unchanged by the permutation). As a result, we shall design the reflection arrays as n rows of linear arrays (Figs 7 and 8), each one (of length n) responsible for reflecting a single row. RL1, the array for on-the-fly reflection of a row, will be used later, as a building block in the design of the arrays for rotation.

The array developed in Sec 5 has the following structure: Each PE has output registers to the left and right. The value from the first PE must move $n - 1$ steps to the last PE. In general, the value in the i-th PE (for $i \leq n/2$) must move $n - 2i + 1$ steps, to PE $n - i + 1$. For $i > n/2$, the PEs must send their data to the *left* in a corresponding manner. We ensure that there is no conflict during data transmission as follows. Initially all PEs are in *idle* mode, where they merely transmit the left input to the right and vice-versa. The first

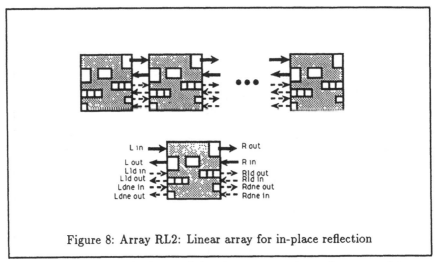

Figure 8: Array RL2: Linear array for in-place reflection

PE, when it receives a *load* control signal, say, at $t = 1$ puts its accumulator on its right output. The second PE cannot do the same until it has successfully transmitted the value sent by the first one, i.e., until $t = 3$. In general, the i-th PE receives the first data value at $t = i$, and must transmit for $i - 2$ more time units until can send its own data. The *load* signal must therefore reach the i-th PE at $t = 2i - 1$, i.e., it moves by one processor every two time units.

As mentioned before the scheme has the flaw that it is not implementable using purely local control: the i-th data value reaches its destination (PE $n - i + 1$) at time $t = $ starting time + distance to destination i.e., $(2i - 1) + (n - 2i + 1) = n$. However, this requires that all the PEs must load their registers from the input data at $t = n$, which implies a broadcast signal.* The solution is to slow down the *load* signal, so that it reaches the i-th PE at $t = 3i - 2$ (as indicated by the *three* stage output buffer in Fig 8). Thus the i-th data value reaches its destination at $t = n + i - 1$, and the termination can be achieved by sending an *unload* signal to the rightmost PE at $t = n$, which travels *left* at the rate of one PE per time unit. The problem of whether a PE is in the right half or the left half does not need to be hardwired, since the load signal that is the first to arrive determines this. The entire reflection is completed in $1.5n$ time steps. It has been shown elsewhere that this time bound is also optimal, given the constraint of no broadcasts. As mentioned above, the array for reflecting the entire matrix merely consists of n rows of such linear arrays.

---

*By our rules, such a broadcast signal is not permitted in a *linear array*. However, if this array is extended to a 2-dimensional one, it is permitted if sent to only a *boundary row*. We could, therefore, use n copies of the linear array with broadcast as a row in the reflection array, by propagating the stop signal *downwards*. This would mean that the first PE in the i-th row starts at $t = i$, yielding an array that takes $2n$ time steps. Our proposed solution is better than this.

6.1 On the fly reflection

In the array RL1, each processor has an accumulator, a data input from the left, and an output to the right, and a control signal that it receives from the left. If the control signal is one (indicated by solid dots in the figure), the processor latches the input value into the accumulator and from the next cycle onwards, it simply transmits the input values to the right. It should be clear that this array reverses the incoming stream of data values in $2n$ time units, as it is being loaded into the array. Note that in Fig 7, the control signal (i.e., the solid dot) is always aligned to the first data value in the stream. This time bound is optimal, since the last value to enter the array (at $t = n$) is $a_{1,1}$, and its destination is PE n.

In order to describe and verify the correctness of this architecture, we need to develop space-time mappings. Static maps to memory addresses cannot describe dynamic behaviour like data flowing over an input line. For instance, the input of the operation is described by the following space-time mapping:

$$(i,j) \mapsto \langle j - 1, (i, n, \mathrm{DT}) \rangle$$

where the address space is $\{0, \ldots, n-1\} \times \{0, \ldots, n\} \times \{\mathrm{Acc}, \mathrm{DT}\}$. (The column $y = n$ is added just to simplify the input description. For the same reason, we allow the time -1 for the first input.) The output is described by the mapping

$$(i,j) \mapsto \langle 2(n - 1), (i, j, \mathrm{Acc}) \rangle.$$

Given a formal transfer relation, describing the data transfers of the array as informally described above, the correctness of the on the fly-reflection can be proved similarly as for the in-place reflection. The details are left as an exercise for the reader.

7 90-degree Rotation of a Stored Matrix

This operation rotates the storage order of a matrix that is already present in the array, as shown in Fig. 9. The array that we design implements the space time tarnsfer relation shown in Fig. 11

Since arithmetic in Eqn 2 is mod-n, $C(i,j) = (j, n + 1 - i)$. We observe that here too, the matrix can be partitioned so that the action of C can be decomposed into *independent* permutations of these partitions. In particular, let $i \leq n/2$ and $m = n - i + 1$, and consider the elements in the set $A_i = A_1 \cup A_2 \cup A_3 \cup A_4$, where $A_1 = \{a_{i,k} \mid k = i \ldots m-1\}$, $A_2 = \{a_{k,m} \mid k = i \ldots m-1\}$, $A_3 = \{a_{m,k} \mid k = m - 1 \ldots i + 1\}$, and $A_4 = \{a_{k,i} \mid k = m \ldots i + 1\}$. Then $C(A_i) = A_i$, i.e., A_i is closed under the rotation, C. The set A_i consists of a "square ring," on which the action of C is merely a circular right shift by $m - i - 1$, i.e., by $n - 2i + 1$. In our array, therefore, if we design the PEs to align themselves into $n/2$ circular shift-registers, there will be no data conflicts. Moreover, the i-th such shift-register will need to perform $n - 2i + 1$ shifts. This can be achieved as follows. Let us first consider PEs that are in the interior of the triangle $[1,1]$, $[1,n]$ and $[n/2,n/2]$ (we call such PEs *north quadrant* PEs). We see that if a PE in the i-th row starts at $t = i$ and stops at $t = n - i + 1$, it will perform the desired number of shifts. This can be signalled by a *start* signal that propagates downward (at unit speed), and a *stop* signal moving upwards from the bottom of the array. However, for the PEs in the south quadrant, the same signals can be used simply by interchanging their roles — the signal going up is treated as the *start* signal and the one going down is the *stop* signal. The PEs in the east and west quadrant are similar, except that they perform vertical shifts. As with the reflection array, we do not need to hardwire any of the PEs, the first control signal to reach a PE indicates which quadrant it is in. For example, if a PE first receives a 1 on its bottom imput, it knows that it is in the south quadrant and must therefore shift left until it gets a signal from the top. Moreover, the corner PEs are also uniquely informed of their duties because they are the ones that get *two* control signals simultaneously. The complete array is shown in Fig 10 (for the sake of clarity, we do not show the data connections that are not used). The entire operation takes n time steps, which is optimal.

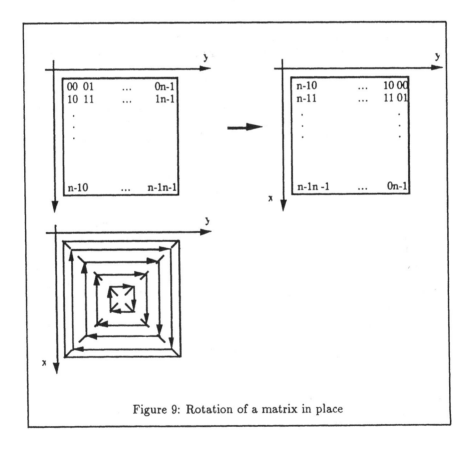

Figure 9: Rotation of a matrix in place

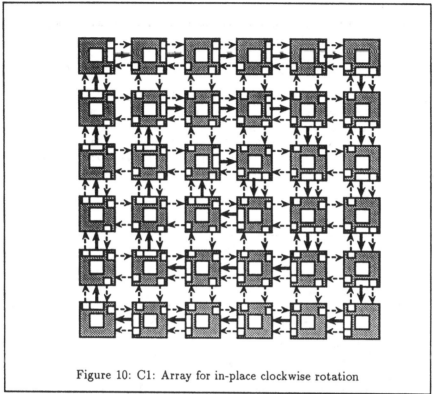

Figure 10: C1: Array for in-place clockwise rotation

8　90 degree rotation of an incoming matrix

This operation rotates the storage order of a matrix on the fly, while it is loaded into the mesh array. The array is loaded from the "right", column by column, with the lowest-numbered column first. When the matrix eventually is loaded it has become stored in a rotated fashion, so that column j becomes row j and row i becomes row $n - 1 - i$. See figure 12.

Let us now see how we can develop an array for performing a rotation of a matrix as it is being loaded into the array. We know that the i-th row of the matrix (elements in the order $a_{i,n}, a_{i,n-1}, \ldots, a_{i,2}, a_{i,1}$) enters the array from the left, and their destination is the $(n - i + 1)$-th column. We can send the values to the correct destination if the entire row is first transmitted to the right, and when, at $t = n - i + 1$, it reaches PE $[i, n - i + 1]$ (the *pivot* PE), it is sent on down for another $n - i + 1$ time units. At this time, all the PEs below (inclusive) the "south-west-north-east" diagonal latch their input values into their accumulators, the pivot PEs start routing the data upwards, and the PEs above the diagonal act like a vertical liear array for reflection for another $2(i - 1)$ steps. At this time the operation is complete.

We must now ensure that the above sequence of operations can be performed

with only local control. Initially all the PEs are idle, and they start to transmit
data to the right when they recieve a control signal from the left (this control
signal travels to the right at unit speed, and is thus always aligned with $a_{i,n}$.
There is also a vertical control signal that travels upwards at unit speed. In
addition, there is a pair of slow control signals that move (one vertically, and
the other horizontally) at half speed. All signals are initiallized to one at $t = 1$.
Now, the pivot PEs are the ones that recieve both the fast control signals
simultaneously. Bottom PEs are the ones that get the vertical (fast) signal
first. These PEs ignore the vertical slow signal (merely transmit it). They
start shifting downwards as soon as they get the horizontal fast signal, and
load their accumulators from the vertical input when the horizontal slow signal
arrives. The pivot PEs atart shifting (Left-input to Down-output) when they
get the fast signal, load their accumulators when they get the slow signal, and
from then on start transmitting upwards (Left-Input to Up-output). The top

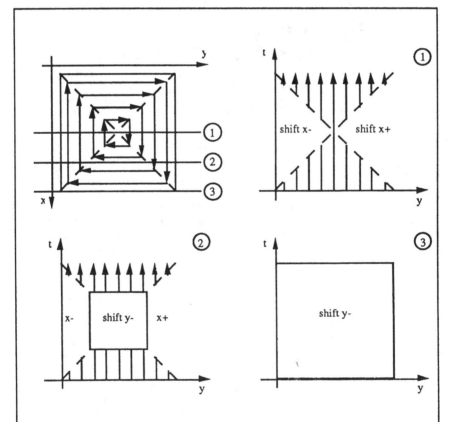

Figure 11: Space time view of the transfer relation for reflection: three diferent
slices of the array are shown

half PEs remain idle until they get the vertical slow signal, and they use it as the control signal to perform a vertical reversal of the upward-moving data stream.

The reader may use a formal argument like the ones presented earlier to prove the correctness of the array. The array takes $2n$ time steps to complete the operation, and once again, a critical path argument shows that this is optimal.

9 Conclusions

We have described how to perform affine permutations of matrices on mesh-connected processor arrays with local control only. The data permutations can be used to interface matrix computations on systolic and wavefront array architectures.

First, we described how to implement general affine permutations by an efficient on the fly-generation of tags, while loading the matrix to be permuted, followed by a sorting phase. The generality here was bought at the expense of

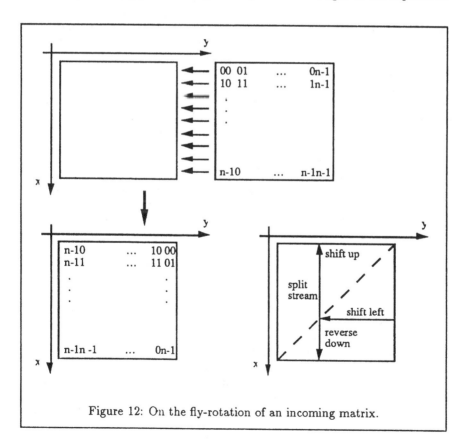

Figure 12: On the fly-rotation of an incoming matrix.

providing hardware for generating the tags and implementing a systolic sorting algorithm on the mesh. The sorting phase also adds to the time for the
operation.

In order to alleviate these problems, we devised four specialized architectures: two for in-place reflection and 90 degree rotation, and two for the same
operations performed on the fly while loading the matrix into the mesh. With
these operations, any affine permutation in the subgroup of 90 degree rotations
and reflections can be carried out. This subgroup includes matrix transpose.
The architectures for in-place rotation and the on the fly-permutations are
time-optimal even without the restriction on local control. There are faster
architectures for in-place reflection with broadcast signals. These schemes require, however, that all cells at some time change their behaviour simultaneously. If a cell is to change its behaviour only at the arrival of a control signal,
then the proposed architecture is time-optimal.

In the proposed architectures, cells behave differently depending on their
position. They set their state according to their position by monitoring which
control signal arrives first. Actually, control signals input over the boundaries
and propagated with different speed provide a general way to divide the mesh
into regions, where different signals arrive first in different regions. By choosing
input boundaries and varying the input times and the speed of the signals, any
partitioning of the cells into regions with linear bounds can be effectuated. In
this way the states of the cells can be set "remotely", from the boundaries of
the mesh, according to their position.

The four proposed permutation architectures could be seen as active memory operations. it is indeed possible to build such a memory, with local control
only, as a mesh-connected array of a single, simple type of cells. Such a memory
could serve as a frame buffer for systolic arrays. Suitable rotations and reflections could then be performed on the buffer before the systolic computation
is initialized. Alternatively, the operations could be carried out in the systolic
array itself, if it is mesh-connected.

A theoretical framework for describing and reasoning about prescheduled
data transfers was developed. An immediate application of the framework is
verification of the proposed specialized architectures, but the framework is general and can describe other topologies and other operations than permutations
as well. The introduction of address space-time adds the power to describe
both temporal and spatial distributions of data, which is absolutely crucial
when reasoning about systolic and wavefront array architectures which operate
in a heavily pipelined fashion.

Automated synthesis and verification of implementations of space-time permutations is an intriguing possibility. In the case of mesh-connected arrays with
local control only, we believe that there are efficient procedures for this. In such
a system, data travels along straight lines in given registers in address space-
time. To show that two events are connected then amounts to solving a number
of linear equations. The transfer relations will typically be compactly represented by linear inequalities, deciding where in space-time certain behaviours
are to occur. Such inequalities can be reasoned about efficiently. Fast methods

exist to decide whether or not a system of linear inequalities has a solution [Sho81]. Thus, it is for instance possible to decide whether a representation of a transfer relation really is well-formed, i.e. represents a causal forest.

It is also interesting to consider the relationships of the transfer relation – space-time permutation model with space-time scheduling methods for synthesis of regular synchronous hardware. The specialized mesh-connected architectures considered here can be described by systems of uniform recurrence equations [KMW67]. Algorithms described by such equations can be automatically mapped to systolic arrays [Qui84]. They are included in a larger class of static algorithms, for which methods to find time-optimal mappings under linear constraints exist [Lis90]. Thus, it seems viable that these transformational methods can be used to optimize mesh-connected permutation architectures with localized control as well.

The theoretical framework is by no means restricted to systems with local control: the actions of other types of systems like SIMD machines can be also be described. Verification of the correctness of SIMD data distribution operations could, for instance, be done by giving a transfer relation as formal semantics to each data-moving instruction, combining transfer relations arising from sequences of such instructions and then verifying that the resulting transfer relation connects the desired events.

References

[CF85]   K. Culik II and I. Fris. Topological transformations as a tool in the design of systolic networks. *Theoretical Computer Science*, 37:183–216, 1985.

[Che86]   Marina C. Chen. A design methodology for synthesizing parallel algorithms and architectures. *Journal of Parallel and Distributed Computing*, 3(6):461–491, December 1986.

[CS84]   Peter R. Cappello and Kenneth Steiglitz. Unifying VLSI designs with linear transformations of space-time. *Advances in Computing Research*, 2:23–65, 1984.

[DI87]   Jean-Marc Delosme and Ilse Ipsen. Efficient systolic arrays for the solution of Toeplitz systems: an illustration of a methodology for the construction of systolic architectures in VLSI. In Will Moore, Andrew McCabe, and Roddy Urquhart, editors, *Systolic Arrays*, pages 37–46, Bristol, UK, 1987. Adam Hilger.

[Fla82]   Peter M. Flanders. A unified approach to a class of data movements on an array processor. *IEEE Transactions on Computers*, C-31(9):809–819, September 1982.

[HL87]   Chua-Huang Huang and Christian Lengauer. The derivation of systolic implementations of programs. *Acta Informatica*, 24(6):595–632, November 1987.

[JRK87]   H. V. Jagadish, Sailesh Rao, and Thomas Kailath. Array architec-
          tures for iterative algorithms. *Proceedings of the IEEE*, 75(9):1304–
          1321, September 1987.

[KL80]   H. T. Kung and C. E. Leiserson. *Algorithms for VLSI Processor
          Arrays*, chapter 8.3, 'Introduction to VLSI Systems,' Mead, C. and
          Conway, L., pages 271–292. Addison-Wesley, Reading, Ma, 1980.

[KMW67]   R. M. Karp, R. E. Miller, and S. Winograd. The organization of
          computations for uniform recurrence equations. *JACM*, 14(3):563–
          590, July 1967.

[Kun88]   S. Y. Kung. *VLSI Array Processors*. Prentice Hall, 1988.

[Lis89]   Björn Lisper. *Synthesis of Synchronous Systems by Static Schedul-
          ing in Space-time*, volume 362 of *Lecture Notes in Computer Sci-
          ence*. Springer-Verlag, Heidelberg, May 1989.

[Lis90]   Björn Lisper. Synthesis of time-optimal systolic arrays with cells
          with inner structure. *Journal of Parallel and Distributed Comput-
          ing*, 10(2):182–187, oct 1990.

[LMT]   Tom Leighton, Filia Makedon, and Ioannis Tollis. A $2n − 2$ step
          algorithm for routing in an $n \times n$ array with constant size queues.
          Manuscript, Laboratory for Computer Science, MIT, Cambridge,
          MA 02139.

[LR]   Björn Lisper and Sanjay Rajopadhye. Prescheduled permutations
          of distributed data fields. Manuscript in preparation.

[MF86]   Dan I. Moldovan and Jose A. B. Fortes. Partitioning and map-
          ping algorithms in fixed size systolic arrays. *IEEE Transactions on
          Computers*, C-35:1–12, January 1986.

[MMJ89]   John McCanny, John McWirther, and Earl Schwartzlander Jr., ed-
          itors. *Systolic Array Processors*. Prentice Hall, Hertfordshire, UK,
          1989.

[MMU87]   Will Moore, Andrew McCabe, and Roddy Urquhart, editors. *Sys-
          tolic Arrays*. Adam Hilger, Bristol, UK, 1987.

[Mol82]   Dan I. Moldovan. On the analysis and synthesis of VLSI algorithms.
          IEEE Transactions on Computers, C-31:1121–1126, October 1982.

[MW84]   W. L. Miranker and A. Winkler. Space-time representation of com-
          putational structures. *Computing*, 32:93–114, 1984.

[NS79]   D. Nassimi and S. Sahni. Sorting on a mesh-connected parallel
          computer. *IEEE Transactions on Computers*, C-28(1):2–7, January
          1979.

[NS80] David Nassimi and Sartaj Sahni. An optimal routing algorithm for mesh-connected parallel computers. *Journal of the Association for Computer Machinery*, 27(1):6–29, January 1980.

[O'L87] Dianne P. O'Leary. Systolic arrays for matrix transpose and other reorderings. *IEEE Transactions on Computers*, C-36(1):117–122, January 1987.

[OO87] A. Yavuz Oruç and M. Yaman Oruç. Programming cellular permutation networks through decomposition of symmetric groups. *IEEE Transactions on Computers*, C-36(7):802–809, July 1987.

[Qui84] Patrice Quinton. Automatic synthesis of systolic arrays from uniform recurrent equations. In *Proceedings of the 11th Annual International Symposium on Computer Architecture*, pages 208–214, June 1984.

[Qui87] Patrice Quinton. *The Systematic Design of Systolic Arrays*, chapter 9, Automata Networks in Computer Science, pages 229–260. Princeton University Press, 1987. Preliminary versions appear as IRISA Tech Reports 193 and 216, 1983.

[QV89] Patrice Quinton and Vincent Van Dongen. The mapping of linear recurrence equations on regular arrays. *Journal of VLSI Signal Processing*, 1(2):95–113, 1989.

[Raj89] Sanjay V. Rajopadhye. Synthesizing systolic arrays with control signals from recurrence equations. *Distributed Computing*, pages 88–105, May 1989.

[Rao85] Sailesh Rao. *Regular Iterative Algorithms and their Implementations on Processor Arrays*. PhD thesis, Stanford University, Information Systems Lab., Stanford, Ca, October 1985.

[RF90] Sanjay V. Rajopadhye and Richard M. Fujimoto. Synthesizing systolic arrays from recurrence equations. *Parallel Computing*, 14:163–189, June 1990.

[RFS85] I. V. Ramakrishnan, D. S. Fussell, and A. Silberschatz. Mapping homogeneous graphs on linear arrays. *IEEE Transactions on Computers*, C-35:189–209, March 1985.

[RK88] Sailesh K. Rao and Thomas Kailath. Regular iterative algorithms and their implementation on processor arrays. *Proceedings of the IEEE*, 76(3):259–269, March 1988.

[RTRK88] Vawani Roychowdhury, Lothar Thiele, Sailesh K Rao, and Thomas Kailath. On the localization of algorithms for VLSI processor arrays. In Robert W. Brodersen and Howard S. Moscovitz, editors,

VLSI Signal Processing, III, pages 459–470, Monterey, Ca, November 1988. IEEE Accoustics, Speech and Signal Processing Society, IEEE Press. A detailed version is submitted to IEEE Transactions on Computers.

[Sho81]   Robert Shostak. Deciding linear equalities by computing loop residues. *Journal of the Association for Computer Machinery*, 28(4):769–779, October 1981.

[Sie85]   Howard Jay Siegel. *Interconnection Networks for Large-Scale Parallel Processing*. Lexington Books, Lexington, MA/Toronto, 1985.

[TK77]   Clark D. Thompson and H. T. Kung. Sorting on a mesh-connected parallel computer. *CACM*, 20(4):263–271, April 1977.

[WAS88]   Benjamin W. Wah, Mokhtar Aboelaze, and Weijia Shang. Systematic design of buffers in macropipelines of systolic arrays. *Journal of Parallel and Distributed Computing*, 5(1):1–25, February 1988.

[WD89]   Yiwan Wong and Jean Marc Delsome. Transformation of broadcasts into propagations in systolic algorithms. Technical Report YALEU/DCS/RR-701, Yale University, Computer Science Department, May 1989.

[YC88]   Yoav Yaacoby and Peter R. Cappello. Converting affine recurrence equations to quasi-uniform recurrence equations. In *AWOC 1988: Third International Workshop on Parallel Computation and VLSI Theory*. Springer Verlag, June 1988. See also, UCSB Technical Report TRCS87-18, February 1988.

7

ARCHITECTURES FOR
STATICALLY SCHEDULED DATAFLOW[1]

Edward Ashford Lee
Jeffrey C. Bier

U. C. Berkeley
Berkeley, CA 94720

ABSTRACT

When dataflow program graphs can be statically scheduled, little run-time overhead (software or hardware) is necessary. This paper describes a class of parallel architectures consisting of Von Neumann processors and one or more shared memories, where the order of shared-memory accesses is determined at compile time and enforced at run time. The architecture is extremely lean in hardware, yet for a set of important applications it can perform as well as any shared memory architecture. Dataflow graphs can be mapped onto it statically. Furthermore, it supports shared data structures without the run time overhead of I-structures. A software environment has been constructed that automatically maps signal processing applications onto a simulation of such an architecture, where the architecture is implemented using Motorola DSP96002 microcomputers.

Static (compile-time) scheduling is possible for a subclass of dataflow program graphs where the firing pattern of actors is data-independent. This model is suitable for digital signal processing and some other scientific computation. It supports recurrences, manifest iteration, and conditional assignment. However, it does not support true recursion, data-dependent iteration, or conditional evaluation. An effort is under way to weaken the constraints of the model and to determine the implications on hardware design.

---

[1] Reprinted with permission from *Journal on Parallel and Distributed Systems*, December 1990. The authors gratefully acknowledge support from Darpa, the Semiconductor Research Corporation, Motorola, Inc., and Dolby Laboratories.

1. INTRODUCTION

A promising alternative to general-purpose parallel computing paradigms is to collect a heterogeneous set of special-purpose solutions. Hardware costs have dropped sufficiently that specialized hardware modules can form part of a cost-effective high-performance system. This paper describes a parallel architecture that supports a class of algorithms with reasonably deterministic behavior. The architecture is simple to implement, yet very fast for the class of applications supported. Furthermore, since the class of applications is restricted, fully automated software mapping is feasible.

Among the applications of this architecture are digital signal processing (DSP) and some scientific computing. These applications differ from general purpose computation both in the nature of the algorithms and in the target hardware. The algorithms tend to have less decision making and use mostly simple data structures (arrays and streams). In DSP applications, the target hardware is often dedicated to an application or a small class of applications, rather than being general purpose, and often has to have low cost together with very high computation rates (to meet hard real-time constraints). These differences create both a hindrance and an opportunity. The hindrance is that mainstream computer science techniques do not apply very well. This accounts for the fact that the DSP community designs its own microprocessors, computer languages, multiprocessor architectures, and software. The opportunity is that the structural simplicity of the algorithms and data structures makes some traditionally very difficult problems much easier.

Dataflow techniques have been applied to DSP in the guise of "block-diagram languages" since its very earliest days. Whereas most of the computer user community resists the introduction of new programming paradigms, the DSP community has embraced experimentation of this type. Dataflow representation of algorithms, in fact, is very natural in DSP, appealing *even without the motivation of concurrency*. Of course, the ability to automatically exploit concurrency can only increase the appeal. However, most attempts to do so through the use of dataflow architectures have not succeeded commercially. I propose in this paper that the principal reason for this is that the dataflow techniques of general-purpose computing are too expensive for DSP and more powerful than what is required. The focus of this discussion is on scheduling, the heart of concurrency in dataflow.

In the process of developing a general scheduling strategy suitable for DSP, we have to be realistic in our assertions about the algorithms; specifically, almost any generalization has counterexamples. Although we can rely on *relatively little* decision making in the algorithms, we cannot rely on *no* decision making without sacrificing a great number of applications. Consequently, the proposed scheduling strategies tolerate decision making, but the performance may degrade as the amount of decision making increases. This tolerance, however, means that elements of the strategy may be applicable in general-purpose computing.

1.1. A Scheduling Taxonomy

In this paper, we only consider non-preemptive scheduling, and the emphasis will be on practical solutions rather than unrealistic abstract models. For the purposes of this paper, we define "scheduling" to include three tasks: (1) assigning actors to processors, (2) ordering the actors on each processor, and (3) specifying their firing time. *Every* dataflow implementation must perform all three tasks, but implementations can differ by performing them at compile time or at run-time, or by using complex or simple scheduling strategies. Depending on which tasks are done when, we define four classes of scheduling. The first is *fully-dynamic*, where actors are scheduled at run-time only. When all input operands for a given actor are available, the actor is assigned to an idle processor. The second type is *static allocation*, where an actor is assigned to a processor at compile time and a local run-time scheduler invokes actors assigned to the processor. In the third type of scheduling, the compiler determines the order in which actors fire on each processor. At run-time, each processor waits for data to be available for the next actor in its ordered list, and then fires that actor. We call this *self-timed* scheduling because of its similarity to self-timed circuits. The fourth type of scheduling is *fully-static*; here the compiler determines the exact firing time of actors, as well as their assignment and ordering. This is analogous to synchronous circuits. As with most taxonomies, the boundaries between these categories are not rigid.

1.2. Examples

We can give familiar examples of each of the four strategies applied in practice. Fully-dynamic scheduling has been applied in the MIT static dataflow architecture [Den80], the LAU system, from the Department of Computer Science, ONERA/CERT, France [Pla76], and the DDM1 [Dav78]. It has also been applied in a digital signal processing context for coding vector processors, where the parallelism is of a fundamentally different nature than that in dataflow machines [Kun87]. A machine that has a mixture of fully-dynamic and static-assignment scheduling is the Manchester dataflow machine [Wat82]. Here, groups of 15 processing elements are collected in rings. Actors are assigned to a ring at compile time, but to a PE within the ring at run time. Thus, assignment is dynamic within rings, but static across rings.

Examples of static-assignment scheduling include many dataflow machines. A commonly adopted practical compromise in these machines is to allocate the actors to processors at compile time. Many implementations are based on the tagged-token concept [Arv82]; for example TI's data-driven processor (DDP) executes Fortran programs that are translated into dataflow graphs by a compiler [Cor79] using static-assignment. Another example (targeted at digital signal processing) is the NEC uPD7281 [Cha84]. The cost of implementing tagged-token architectures has recently been reduced significantly using an "explicit token store" [Pap88]. Another example of an architecture that assumes static-assignment is the proposed "argument-fetching dataflow architecture" [Gao88], which is based on the argument-fetching data-driven principle of Dennis and Gao [Den88].

When there is no hardware support for scheduling (except perhaps synchronization primitives), then self-timed scheduling is usually used. Hence, most applications of today's general-purpose multiprocessor systems use some form of self-timed scheduling, using for example CSP principles [Hoa78] for synchronization. In these cases, it is often up to the programmer, with meager help from a compiler, to perform the scheduling. A more automated class of self-timed schedulers targets wavefront arrays [Kun88]. Taking a broad view, asynchronous digital circuits can also be said to use self-timed scheduling.

Systolic arrays, SIMD (single instruction, multiple data), and VLIW (very large instruction word) computations [Fis84] are fully-statically scheduled. Again taking a broad view of the meaning of parallel computation, synchronous digital circuits can also be said to be fully-statically scheduled.

1.3. Generality

As we move from scheduling strategy number one to strategy number four, the compiler requires increasing information about the actors in order to construct good schedules. However, assuming that the information is available, the ability to construct deterministically optimal schedules increases. To construct an optimal fully-static schedule, the execution time of each actor has to be known; This requires that a program have only deterministic and data-independent behavior. Constructs such as conditionals, data-dependent iteration, and some recursion make this impossible, and realistic I/O behavior makes it impractical.

Self-timed scheduling in its pure form is effective for only the subclass of applications where there is no data-dependent firing of actors, and the execution times of actors do not vary greatly. However, unlike fully-static scheduling, some variation in execution times is tolerable. Signal processing algorithms and scientific computation often fit this model. The run-time overhead is very low, consisting only of simple handshaking mechanisms, and requiring no sophisticated hardware capability such as indivisible "fetch-and-add" or "fetch-and-set" primitives. Furthermore, provably optimal (or close to optimal) schedules can be found. As with fully-static scheduling, conditionals, data-dependent iteration, and recursion are excluded if the resulting schedule must be optimal.

Static-assignment scheduling is a compromise that admits more data dependencies than either fully-static or self-timed, but all hope of optimality must be abandoned in most cases. Although static-assignment scheduling is commonly used, compiler strategies for accomplishing the assignment are not satisfactory. Numerous authors have proposed techniques that compromise between interprocessor communication cost and load balance [Muh87] [Chu80] [Zis87] [Ma82] [Efe82] [Lu86]. But none of these consider precedence relations between actors. To compensate for ignoring the precedence relations, some researchers propose a dynamic load balancing scheme at run-time [Kel84][Bur81][Iqb86]. Unfortunately, the cost of such a scheme can be nearly as high as fully dynamic scheduling. Others have attempted with limited success to incorporate precedence information in heuristic scheduling strategies. For instance, Chu and Lan use very simple stochastic computation models to derive some principles that can guide heuristic assignment for more general computations [Chu87]. However, only very simple stochastic models yield to analysis, so we should not expect too much from the resulting principles.

Fully-dynamic scheduling is most able to utilize resources and to fully exploit the concurrency of a dataflow representation of an algorithm, regardless of the amount of data dependency. However, it requires too much hardware and/or software run-time overhead. For instance, the MIT static dataflow machine [Den80] proposes an expensive broadband packet switch for instruction delivery and scheduling. Furthermore, it is usually not practical to make globally optimal scheduling decisions at run-time, so practical implementations fall short of the theoretical ability to exploit parallelism. One attempt to overcome this by using compile-time information to assign priorities to actors to assist a dynamic scheduler was rejected by Granski et. al., who conclude that there is usually not enough performance improvement to justify the cost of the technique [Gra87]. However, in the special case of algorithms with "regular, static structure" (such as a DFT), there are significant performance improvements. But it is precisely such algorithms that require *only* static scheduling, and can therefore be efficiently implemented on much less expensive machines that include no runtime mechanism for scheduling actors. A cost effective solution for a general-purpose computer might be an architecture that can revert to imperative control flow (or something resembling it) when executing algorithms that are statically scheduled. Perhaps some of the recently proposed hybrid von Neumann/dataflow architectures could take advantage of this observation (see for example [Nik89][Ian88]).

In view of the high cost of fully-dynamic scheduling, static-assignment and self-timed are attractive alternatives. This is true even though both will suffer in performance, compared to fully-dynamic scheduling, as the amount of data dependency increases. Self-timed is more attractive for scientific computation and digital signal processing, because it is more static, while static-assignment may be more attractive when there is more data dependency. The performance of both techniques depends heavily on good compile-time decisions, so it is appropriate to concentrate on finding good compiler algorithms.

1.4. Strategy

For any scheduling strategy that requires compile-time decisions, for example assignment or ordering, these decisions can be made by constructing a fully-static schedule and discarding the information that is not required. At run time, the execution is forced to exactly match the "retained" information. For example, in static-assignment, only the assignment information is retained. In self-timed, the assignment and ordering information is retained. In fully-static scheduling, all information is retained.

We will explore a model lying between fully-static and self-timed, and develop an approach to architecture design well matched to this model. In this new model, we retain not only the ordering of actors on each processor, but also the ordering of accesses to shared resources, such as shared memory or shared data structures. Architectures supporting this model are only slightly more complex than architectures supporting fully-static scheduling, but the model is much more robust. Specifically, some of the flexibility of self-timed scheduling persists, because timing information in the fully-static schedule is discarded. Hence the execution time of actors can vary at run time without affecting the correctness of the execution. On the other hand, the run-time execution is more constrained than for self-timed, because the order in which processors access shared resources is forced at run time to exactly match that of the fully-static schedule.

2. SHARED MEMORY ARCHITECTURES

For fully-static implementations the target architecture need not have any special hardware for run-time scheduling. For self-timed scheduling, the only additional requirement is efficient handshaking. In both cases, Von Neumann processing elements are adequate; there is no need to resort to dataflow machines. This section explores these advantages by discussing the run-time cost of three architectures, two designed for self-timed scheduling, and one designed for a new model lying between self-timed and fully-static scheduling. The first two require semaphore-based synchronization, implemented in software and hardware. The third does not require semaphores.

We assume a host carries out the compilation, mapping an application program onto parallel processors that run under control of the host. The parallel processors are designed for signal processing and scientific computing, so we will not be so ambitious as to try to map the compiler and operating system onto the same set of parallel processors by the same techniques. However, with the dropping cost of hardware, a heterogeneous multiprocessor system of this type is attractive. Different parts of the system are specialized to different functions, and hence can do a much better job than a completely "general-purpose" solution.

2.1. Software Semaphores

At Berkeley, we have implemented a limited dataflow programming environment (Gabriel) for digital signal processing that targets multiprocessor machines made with programmable DSP microprocessors [Lee89b]. In this case, the compiler and scheduler produce assembly code for each Von Neumann processor in the system. Self-timed scheduling can be used, so there are no dataflow principles invoked at run-time, except that semaphore-based synchronization is used when tokens pass between processors.

Gabriel uses macro dataflow actors, where each actor is defined in a more conventional language (Lisp, C or assembly code). The granularity of actors is arbitrary, and tends to be highly variable. We have collected libraries of higher-level actors for popular signal processing algorithms, such as FFTs and digital filters. A typical application consists of tens or hundreds of actors (not thousands) so heuristic scheduling strategies with complexity of order N^2 are usable.

One of the target architectures in our lab, donated by Dolby Laboratories of San Francisco, has four Motorola DSP56000 processors, each with a private memory, plus a single shared memory. Accessing the shared memory requires first requesting the bus, then reading the memory, checking a semaphore, and resetting the semaphore. An ideal transaction is illustrated in figure 1 (ideal means that the bus is free when requested, and the semaphores are in their proper state when checked). It is not necessary to have indivisible test-and-set primitives because the static buffering strategy used

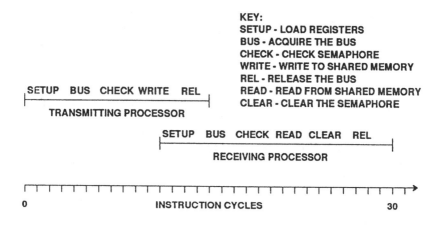

Figure 1. An ideal transaction through shared memory in a typical shared memory multiprocessor when scheduling is self-timed. The number of instruction cycles shown are measured on a prototype multiprocessor system in our Lab.

in Gabriel ensures that no more than one processor will write data to any given memory location, and no more than one processor will read data from that location [Lee87b]. The bus contention/resolution and semaphore handling are the only scheduling overhead incurred at runtime. Nonetheless, in the Dolby architecture these require about 30 instruction cycles for a single transaction, as shown in figure 1. This relatively high cost implies that only large-grain dataflow can be supported efficiently. Furthermore, each token should ideally contain more than a single data value, because the overhead is incurred only once for each token. These restrictions arise from the architecture not being designed with static scheduling in mind. It is much more general than we need.

In programs generated by Gabriel for the Dolby architecture, if the bus is not available when requested, the requesting processor halts until the bus becomes available. Hence contention for the bus can extend the duration of a transaction well beyond the 30 cycles shown in figure 1. In our software implementation of semaphore handling, if the semaphore read from shared memory is not in the desired state, then the processor releases the bus, and attempts the transaction again some time later. The processor busy-waits in the meantime. It is up to the scheduler to ensure that processors do not spend much time busy-waiting. Of course, these repeated reads further increase the load on the shared bus. We conclude that some modification of the architecture is required to be able to effectively map self-timed dataflow graphs onto it.

2.2. Hardware Semaphores

One way to reduce the total transaction time illustrated in figure 1 would be to add hardware that performs the functions we previously performed in software, such as semaphore management. An architecture doing this might look like that in figure 2. Shared memory accesses begin by supplying an address in the shared memory space to the *gate keeper*. The gate keeper asserts a wait signal until the memory request can be satisfied, causing the processor to halt. Meanwhile, the gate keeper acquires the bus, accesses the shared memory, and checks the semaphore. Just as in the software implementation, if the semaphore is not in the desired state, then the read is repeated some time later, with the processor held in its suspended state in the meantime. As before, it is up to the scheduler to ensure that not much time is wasted this way. Since the schedule is self-timed, there is no danger of introducing deadlock by this mechanism.

On writes to shared memory, the gate-keeper need not halt the processor requesting the write. The processor can proceed with its execution until the next shared-memory transaction is encountered. In the meantime, the gate keeper performs the shared-memory write in parallel.

The main advantage of this architecture is that the functions we previously performed in software are performed in hardware, and therefore presumably occur much faster. Furthermore, the processing elements can access shared-memory locations the same way they access local memory.

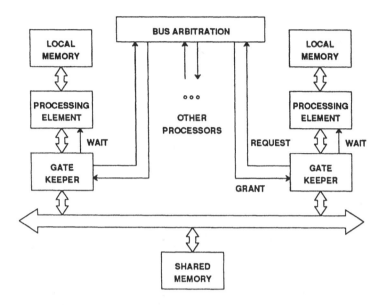

Figure 2. A gated-shared-memory architecture.

The contention resolution and synchronization functions are transparent. However, the gate keeper is not trivial, so the time required for shared-memory accesses is still likely to be larger than the time for local memory accesses, even when there is no contention and the semaphores are in the desired state. In addition, a preliminary design indicates that to implement it on a single chip requires a large (albeit manageable) number of pins. This preliminary design assumed the processing elements would be Motorola DSP96002's, which have separate 32-bit address and data busses (two of each). The relatively high system cost means that it is worth considering an alternative that brings us closer to fully-static scheduling.

2.3. Ordered Memory Architectures

Consider an architecture, shown in figure 3, where a shared bus is not requested by the processors, but rather a central controller (labeled MOMA) grants the bus to processors in some prespecified order (scalability will be addressed shortly). Once the bus has been granted to a processor, it is not released until the processor has completed a shared-memory transaction. The key idea is that a fully-static scheduler can determine *a-priori* the order in which shared-memory transactions will occur. Hence, at the same time that a program is loaded into the private memories of each of the processors, a list is loaded into the controller specifying the order of memory transactions. This list is simply a list of processor numbers, and the controller simply asserts the bus-grant line for each processor in turn. The key advantage

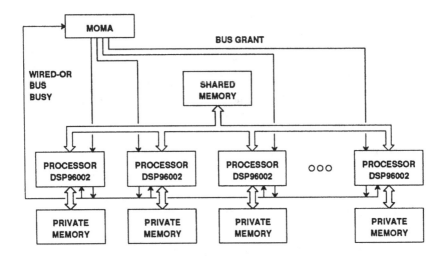

Figure 3. Shared memory accesses can be made extremely efficient when scheduling is static. Here, a controller (MOMA: Maintains Ordered Memory Accesses), grants access to the shared memory in the order predicted by the scheduler. No software or hardware overhead is required for contention/resolution or semaphore handling.

is that *no explicit hardware or software is required for contention/resolution or semaphore management.* Contention is avoided by granting the bus to only one processor at a time. If any processor reaches a code segment where it tries to access shared memory but has not been granted the bus, it simply halts until the bus is granted. Static scheduling ensures that this will not cause deadlock. Static scheduling also ensures that processors do not spend much time halted if there is useful work they could be doing. Furthermore, semaphore synchronization is no longer required. To see this, suppose processor 1 wishes to read a location written by processor 2. Static ordering of memory accesses ensures that the read and write occur in the proper order.

Consider the ideal scenario, where an optimal (minimum makespan) static schedule has been constructed for a completely deterministic program. If any processor halts because it has not been granted the bus, no productivity is lost because no useful work could be done anyway, by the assumption that the schedule is optimal. Stalling the processor is necessary due to precedence constraints. Latency of shared-memory accesses are minimized. In this idealized scenario the function of the central controller is simply to stall processors until their shared memory transactions can proceed. In a more realistic scenario, the function of the central controller is much more important, since it implicitly synchronizes the processors.

The processors in figure 3 are specified as DSP96002's, the latest generation of 32-bit floating-point programmable digital signal processors from

Motorola. This device is well suited to general scientific computation, graphics, and signal processing. More importantly, it is well matched to the proposed architecture because it has two completely independent tri-statable memory busses. It can be wired exactly as shown in figure 3, with no extra logic. The result is a very lean multiprocessor system. We have implemented a four processor prototype (in a software simulation, since the DSP96002 has not been available) using the Frigg hardware simulation environment [Bie89]. These simulations have been used to execute real programs, generated from large-grain dataflow graphs by Gabriel [Lee89b]. In section 3 below, we discuss this software environment in more detail. But for now it is sufficient to observe that as expected, when the scheduler has accurate information to work with, up to 15 times as many shared-memory transactions can be accomplished in a given period of time, compared to the Dolby architecture discussed earlier. Further, we have found that the hardware required to implement an ordered memory architecture is substantially simpler than that required for the implementation of more conventional tightly-coupled multiprocessor architectures.

There is nothing *architecturally* sophisticated in figure 3. The sophistication comes with binding the architecture to a software methodology, discussed in more detail in section 3 below.

2.3.1. Scalability

In general, centralized controllers in multiprocessor systems are ill-advised because they limit scalability. However, we believe that the architecture in figure 3 can be used to find the true limits of scalability of shared-memory architectures. The controller will not be the performance bottleneck; instead, shared-memory bandwidth will be the bottleneck. Unlike many centralized controllers, this one is simple, and can be implemented easily on a single semi-custom chip. A RAM and a very simple finite-state machine are sufficient [Bie90]. The bus grant lines require one pin per processor, in the simplest implementation, so pin count is not a serious problem until hundreds of processors are used together (unlikely even in an efficient shared-memory system). Even if this proves feasible, the bus grant signal could be encoded, and a small number of decoder chips would have to be added. The bus release is accomplished with a single wired-or release signal, so only one pin on the controller is required for this function.

By using the bus arbitration protocol of the 96002, the ordered memory system can be designed so that the central controller does not limit the available shared-bus bandwidth. In the context of the ordered memory architecture, the 96002 bus arbitration protocol functions as follows. The MOMA controller grants shared bus ownership to a processor by asserting that processor's Bus Grant input. When the processor is ready to begin its shared-memory access, it checks the wired-or Bus Busy outputs of its peer processors to ensure that the shared bus is free. If the bus is free, the processor asserts its Bus Busy output and begins its access. Detecting that the processor has begun its shared-memory transaction, the MOMA controller can deassert the processor's Bus Grant input, and assert that of the next processor. When the first processor completes its access, and deasserts its Bus Busy output, the second processor begins its access immediately. Thus, ownership of the shared bus can change hands at the maximum rate permitted by the processor's bus arbitration protocol. The MOMA controller does not introduce any delay.

The bandwidth of the shared memory and bus, however, is likely to become a bottleneck as the number of processors increases, just as with any shared-memory architecture. However, the problem is not as severe as with many traditional shared memory architectures. In the Dolby architecture, for instance, the bus is held for several cycles on each transaction, while a semaphore is fetched and tested and data is read or written. With the ordered-memory architecture, transactions take as little as one cycle, if the scheduler has been successful, because there is no need for semaphore management.

Even with efficient transactions, a shared bus and shared memory will become a problem as the number of processors increases. One solution, first proposed in [Bie90], is shown in figure 4. Two processor clusters have their own MOMA controllers and are connected by a gateway processor. The gateway processor can apply exactly the same ordered-access principle. When a program is loaded into the overall machine, a gateway program is loaded into the gateway processor. This gateway program is trivial, consisting only of a sequence of external memory read and write transactions with the two shared memories. The order of these transactions is enforced by the linear control flow of the program, and the pacing of the program is enforced by the external bus-grant signals. Again, no semaphores are required. Internal memory or registers can be used to buffer data, if necessary. This will be necessary, for example, if several consecutive transactions are with the same shared memory.

In figure 4, the same processor type is being used for the gateway as for other processors, but if its only function is to serve as a gateway, then a simpler custom part could be used. Alternatively, if gateway functions are not sufficient to keep the processor busy, productive computation can be scheduled. The compiler, by assumption, has the information it needs to do this. The implementation shown in figure 4 does not allow for external private memory on the gateway processor, although a simple modification would correct this, at the expense of additional hardware.

The gateway approach in figure 4 illustrates that multiple order-preserving controllers can cleanly coexist. Of course, as with any gateway architecture, the program should be carefully partitioned to prevent the gateway from becoming the bottleneck. This may not always be possible, so of course this architecture is not universally suitable. However, the architecture is easily extended. Any number of gateways can coexist on a bus, so a tremendous number of processor interconnection topologies can be implemented using the ordered-transaction principle.

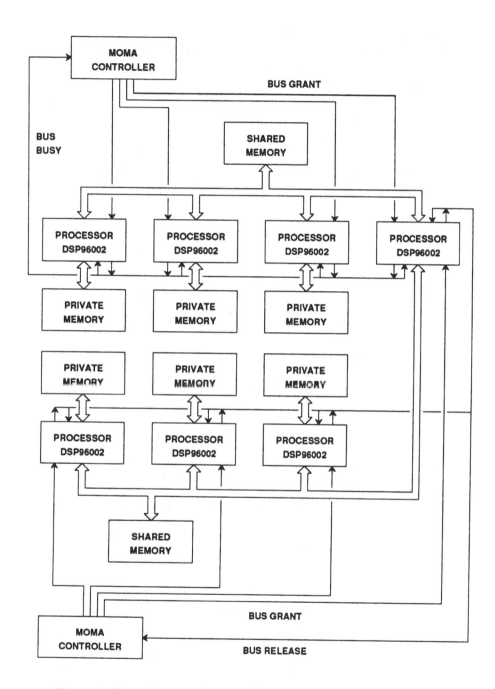

Figure 4. To scale an ordered memory architecture, processor clusters can be connected by a gateway.

2.3.2. Zero Access Time

An interesting enhancement is immediately evident. The static scheduler not only knows when processors will be accessing shared memory, but it knows which memory locations will be accessed, and whether each location will be read or written. This knowledge can be used to speed shared-memory transactions for a relatively slow memory. As it asserts the bus-grant line for a processor, the controller can also drive the address lines of the shared memory, if the shared memory is not in use. If the next transaction is a read, and the processor is not already waiting, then the response time of the memory will appear to the processor to be zero! When the processor gets around to performing the read, the desired data will already be on its data bus.

2.3.3. Estimated Execution Times

One principle limitation with the architecture in figure 3 is the requirement that the order of shared memory transactions be known at compile time. This is possible when fully-static scheduling is possible, which, among other limitations, implies that the execution time of every actor must be known. The Gabriel system does not use fully static scheduling because it is not usually practical to know exactly the execution time of all actors. However, the architecture in figure 3 will function correctly even if *estimates* of the execution time are used. This is because the ordered bus grants provide synchronization, of a sort. If poor estimates of runtimes are used, or a processor is interrupted, then this architecture may yield poorer performance than an architecture that permits dynamic re-ordering of shared-memory accesses, like an architecture that relies on semaphores for synchronization.

Note that if one processor accesses the shared memory at run time much later than the scheduler estimated at compile time, then many (or all) of the other processors might be idled, waiting for bus grants. By contrast, if a processor attempts to access the shared memory too early, only that processor will be idled. This suggests that with more than two processors the scheduler should prefer to overestimate rather than underestimate execution times. This principle is quantified in [Ha89], where a method is given for estimating execution times to minimize expected idle time on all processors. For now, however, we simply assume that the run time estimates are reasonably accurate.

One mechanism that might cause the estimates of actor execution times to be poor is traditional cache management. In many systems, a cache miss can cause a processor to suspend execution of the program for hundreds of cycles. This would probably be unacceptable, since there is a good chance that many of the other processors will be delayed as a consequence. This would not necessarily occur with self-timed scheduling, because shared-memory accesses could be dynamically re-ordered. The same is true of interrupts, and to a lesser degree, data-dependent instruction execution times. Therefore, cache management, interrupt handling, and data-dependent execution-speed optimizations would have to be rethought, or more simply, forbidden. An interesting possible solution is to replace dynamic cache management with static paging. If fully-static scheduling is possible, then in principle, all the requisite information is available. In signal processing, because of hard-real-time constraints, it is common to use software controlled paging rather than dynamic caches [Lee88] [Lee89a].

2.4. Statically Ordered Data Structure Access

The lack of side-effects in dataflow actors makes it particularly difficult to support large, shared data structures [Gau86]. Arvind, et. al., have therefore extended the dataflow model by introducing *I-structures*, a controlled form of global data structures [Arv87]. I-structures are write-once data structures with non-strict semantics, which in practice means that reads may be issued before data is available in the data structure. Support for I-structures requires an ability to queue read requests until they can be satisfied. This mechanism is the most promising available, but it does not come cheaply. One extra memory location is required for each read instruction that can be simultaneously pending. In addition, the I-structure memory needs a processor of a sort to tend to pending reads when a write finally occurs. A much simpler mechanism may be used when scheduling is static.

Consider for example an actor that emits an array. This array might be carried by a single token. Suppose there are two actors that take this array as an argument. A pure dataflow model requires that the array be copied, or at least that an implementation behave as if the array had been copied. Using an I-structure avoids this copying. However, with ordered memory accesses, the copying is not necessary, and neither is the I-structure memory. Since the scheduler is aware of all precedences, it will avoid scheduling reads before the data becomes available. If this cannot be avoided (a processor has nothing to do until the data becomes available), then the read is attempted before the bus is granted to the processor, so the processor halts. The bus will not be granted to the processor until the data is ready. There is no need to queue accesses.

When data passed through the shared memory goes from one actor to one actor, the scheduler can reclaim the token storage after scheduling the read by the destination. Write-once shared data structures are only slightly more complicated, because there may be more than one destination actor. The scheduler can simply use reference counts (RCs) [Hud86][Dri86] to determine when the memory can be reclaimed. For the above example, the RC associated with the array storage would be initialized to 2, the number of destinations, when the array is scheduled to be written. Each time a read is scheduled, the RC is decremented. When it reaches zero the memory can be reclaimed. This works without any run-time overhead because the order of these transactions will be enforced at run-time.

Many variations of this idea immediately come to mind; for example, reference counts could be used for each element of the array, instead of the whole array, thereby getting some of the advantages of the non-strictness of I-structures. Specifically, the array does not have to be completely filled before some of its elements can be read. Also, if the RC of a data structure is identically one, then an actor using it may modify it, instead of simply reading it, something not permitted in the write-once I-structures. An intelligent code generator can get considerable mileage out of this.

The reference count technique has been criticized for a number of reasons [Arv87b], most of which break down when the scheduling is static. Primarily, for ordered memory architectures, the overhead of managing RCs is incurred *only at scheduling time*, not at run time.

3. STATICALLY SCHEDULED CONTROL

The ordered memory architecture and the static shared data structures seem to provide a very clean solution to some vexing problems. However, they are only applicable when fully-static or self-timed scheduling is possible. Although this imposes some serious constraints, the constraints are less serious than they may appear at first.

The programming environment called Gabriel [Lee89b], designed for signal processing applications, is based on graphical dataflow representations of algorithms. Although specialized to signal processing, this environment has permitted extensive experimentation with scheduling algorithms, target architectures, and with a style of programming that matches the need for static scheduling. As mentioned before, we have implemented a software simulation of a four processor ordered memory architecture [Bie90] using the Frigg hardware simulation environment [Bie89], and have retargeted Gabriel to this architecture. Hence, we have been able to gain some experience compiling and running real programs on this architecture.

The granularity of the actors in Gabriel is arbitrary, varying from simple arithmetic operators up to high level signal processing functions such as FFTs. Gabriel translates dataflow graphs into sequential assembly code for programmable DSPs, performing the scheduling statically for multiple processors. A typical signal processing application contains at most hundreds of actors, so we can experiment with rather complex scheduling algorithms without getting bogged down.

To be able to schedule computations statically, Gabriel restricts the dataflow model to a subclass called synchronous dataflow (SDF) [Lee87a]. We begin this section with a review of the properties of this subclass, and then continue by showing that it is not as limited as might first appear. In particular, we show that it supports recurrences, manifest iteration, and conditional assignment, but does not support true recursion, data-dependent iteration, nor conditional evaluation.

3.1. Synchronous Dataflow

A subclass of dataflow graphs lacking data dependency is well suited to static scheduling. Precisely, the term "synchronous dataflow" has been coined to describe graphs that have the following property [Lee87a]:

SDF property:

> A *synchronous actor* produces and consumes a fixed number of tokens on each of a fixed number of input and output paths. An *SDF graph* consists only of synchronous actors.

The basic constraint is that the number of tokens produced or consumed cannot depend on the data. An immediate consequence is that SDF graphs cannot have data-dependent firing of actors, as one might find, for example, in an if-then-else construct. In exchange for this limitation, we gain some powerful analytical and practical properties [Lee87a][Lee87b]:

1) For SDF graphs, the number of firings of each actor can be easily determined at compile time. If the program is non-terminating, as for example in real-time DSP, then a periodic schedule is always possible, and the number of firings of actors within each cycle can be determined at compile time. In either case, knowing these numbers makes it possible to construct a deterministic acyclic precedence graph. If the execution time of each actor is deterministic and known, then the acyclic precedence graph can be used to construct optimal or near-optimal schedules.

2) For non-terminating programs, it is important to verify that memory requirements are bounded. This can be done at compile time for SDF graphs.

3) Starvation conditions, in which a program halts due to deadlock, may not be intentional. For any SDF graph, it can be analytically determined whether deadlock conditions exist.

4) If the execution time of each actor is known, then the maximum execution speed of an SDF graph can be determined at compile time. For terminating programs, this means finding the minimum makespan of a schedule. For non-terminating programs, this means finding the minimum period of a periodic schedule.

5) For any non-terminating SDF graph executing according to a periodic schedule, it is possible to buffer data between actors *statically*. Static buffering means loosely that neither FIFO queues nor dynamically allocated memory are required. More specifically, it means that the compiler can statically associate memory locations with actor firings. These memory locations contain the input data and provide a repository for the output data.

These properties are extremely useful for constructing parallelizing compilers, but they only apply to SDF graphs, and optimal schedules can only be constructed when the execution times of the actors are known. We have been developing techniques that weaken the SDF constraint, thus supporting more general dataflow graphs without resorting to fully dynamic control [Ha89]. However, these techniques require modification of the MOMA controller of the ordered memory architecture. There is still much work to be done to find the best design parameters, so in this paper we will retain the SDF constraint.

Optimal compile-time scheduling of precedence graphs derived from SDF graphs is one of the classic NP-complete scheduling problems. Many simple heuristics have been developed over time, with some very effective ones having complexity n^2, where n is the number of actors (see for example [Hu61]). However, even n^2 complexity can bog down a compiler. Fortunately, the granularity of dataflow actors in Gabriel and the small size of many signal processing applications means that we can ignore this problem for now. To generalize these methods beyond signal processing applications, strategies will probably be needed to cluster sets of actors into macro actors, thus reducing the number of actors to be considered in constructing a schedule. For example, the clustering method proposed in [Kim88] seems suitable.

Static scheduling promises low-cost architectures, at the expense of compile-time complexity. For many applications, this is a very attractive tradeoff. However, only some applications can be statically scheduled. The SDF model, which can be statically scheduled, may appear to lack control constructs because it does not permit data-dependent firing of actors. However, this is not entirely true. Some control structures are possible within SDF, notably recurrences, manifest iteration, and conditional assignment.

3.2. Recurrences

The dataflow community has recognized the importance of supporting *recursion*, or self-referential function calls. To some extent, this ability has become a litmus test for the utility of a dataflow model. The most common implementation, however, dynamically creates and destroys instances of actors. This is clearly going to be problematic for a static scheduler.

In imperative languages, recursion is used to implement recurrences and iteration, usually in combination. If we avoid the notion of "function calls", at least some recurrences can be simply represented as feedback paths in a dataflow program graph. This section will study the representation of recurrences using feedback. This representation poses no difficulty for static scheduling, although to some it lacks the elegance of recursion.

Recurrences depend on the notion of "delays". Once understood, this notion can be used to explain fundamental limits on the concurrency in SDF graphs. It can also be used to relate SDF to static dataflow [Den80]. This is done below.

3.2.1. Delays

A dataflow graph with a recurrence is represented schematically in figure 5. This graph is assumed to fire repeatedly. Borrowing terminology from the signal processing community, the feedback path has a *delay*, indicated with a diamond, which can be implemented simply as an initial token on the arc. A set of delays in a dataflow graph corresponds to a *marking* in Petri nets [Pet77] or to the "D" tag manipulation operator in the U-interpreter [Arv82]. In fact, the symbol "D" was selected to suggest "delay" [Arv89]. A necessary (but not sufficient) condition for avoiding deadlock in an SDF

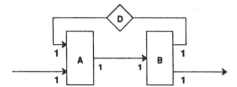

Figure 5. A dataflow graph with a recurrence. Recurrences are expressed using directed loops and delays.

graph is that any directed loop in the graph must have at least one delay.

A *delay* does not correspond to unit time delay, but rather to a single token offset. Such delays are sometimes called *logical delays* or *separators* to distinguish them from time delays [Jag86]. For SDF graphs, a logical delay need not be a run-time operation. Consider for example the feedback arc in figure 5, which has a unit delay. The numbers adjacent to the arcs indicate the number of tokens produced or consumed when the corresponding actor fires. The initial token on the arc means that the corresponding input of actor A has sufficient data, so when a token arrives on its other input, it can fire. The *second* time it fires, it will consume data from the feedback arc that is produced by the *first* firing of actor B. In steady-state, the nth firing of actor B will produce a token that will be consumed by actor A on its $(n + 1)^{th}$ firing; hence the arc has unit token offset. The *value* of the initial token can be set by the programmer, so a delay can be used to initialize a recurrence. When the initial value is other than zero, we will indicate it using the notation D(*value*). Since delays are simply initial conditions on the buffers, they require no run-time overhead. In Gabriel, a delay is a property of an arc in the dataflow graph, rather than being an actor.

3.2.2. Bounds on Performance

Consider non-terminating algorithms, or algorithms that operate on a large data set. For these, directed loops are the only fundamental limitation on the parallelizability of the algorithm. This is intuitive because any algorithm without recurrences can be pipelined. A special case of SDF, called *homogeneous* SDF, is where every actor produces and consumes a single token on each input and output. For homogeneous SDF graphs, it is easy to compute the minimum period at which an actor can be fired. This is called the *iteration period bound,* and is the reciprocal of the maximum computation rate. The iteration period bound may be much smaller than the time required to compute one pass through the dataflow graph. It is a limit on the time per pass if an infinite number of passes are computed.

Let $R(L)$ be the sum of the execution times of the actors in a directed loop L. The iteration period bound is the maximum over all directed loops L of $R(L)/D(L)$, where $D(L)$ is the number of delays in L [Ren81][Coh85]. The directed loop L that yields this maximum is called the *critical loop*. General SDF graphs can be systematically converted to homogeneous SDF graphs for the purpose of computing the iteration period bound [Lee86]. If there are no directed loops in the graph, then we define the iteration period bound to be zero, since in principle all firings of each node could occur simultaneously. It is important to realize that there is nothing fundamental in the following discussion that prevents this. Implementation considerations may make it impractical, however.

Another limitation on concurrency is the notion of state. Particularly in large or medium grain dataflow graphs, it is convenient to permit an actor to remember data from one invocation to the next. This is simply modeled as a self-loop with a unit delay. Such a self-loop precludes multiple simultaneous invocations of the actor, hence this self loop may become the critical loop.

Once the iteration period bound is known, we can derive a bound on the performance of an ordered memory architecture, based on a set of (admittedly) unrealistic assumptions. First, assume we have a completely deterministic dataflow graph, and assume there are enough processors that a hypothetically optimal scheduler can meet the iteration period bound. The iteration period bound does not reflect bandwidth or latency limitations on interprocessor communication, however. For the ordered memory architecture of figure 3, a memory transaction can occur in one cycle of the shared memory. If we assume that the shared-memory cycle time is the same as local memory cycle time[2], then latency adds nothing to the iteration period. Bandwidth limitations, however, may add to the iteration period. Each time the ideal scheduler schedules two simultaneous memory transactions, one of them must be delayed. If one of them is not in the critical path, then that one should be delayed, and there may again be no effect on the iteration period. If both are in the critical path, then the iteration period will be extended by one cycle. If three transactions are scheduled simultaneously, then one of the transactions has to be delayed two cycles, increasing the iteration period by at most two cycles. If M simultaneous transactions are scheduled, then the iteration period increases by at most $M-1$ cycles. If the total number of transactions is T, then the very worst situation increases the iteration bound by at most $T-1$ cycles.

Suppose now that 10 processors are each running a program that accesses shared memory 10% of the time. Then this bound tells us that the ordered shared memory architecture can run this program in at most twice the time of the theoretical minimum. However, this result should not be taken very seriously because the performance will depend much more heavily on scheduling heuristics that are used. The performance can be better (since simultaneous transactions can be studiously avoided [Sih90]) but can also be worse (if, as is likely, a suboptimal scheduling algorithm is

---

[2] This is actually not a bad assumption for the architecture in figure 3, if the number of processors is modest. The main limitation on shared-memory cycle time is likely to be capacitive loading on the shared bus, and the price of the memory, of course.

used). Furthermore, the use of gateways completely undermines this analysis. Thus, this is not a very useful bound.

3.2.3. Bounded Buffer Sizes

Although SDF actors cannot be created at runtime, SDF is not the same as static dataflow [Den80]. For instance, in SDF, there is no impediment to having multiple instances of an actor fire simultaneously, as long as the actor does not have state. A particular implementation, however, may impose such a constraint. Consider for example an implementation that permits no more than one memory location to be associated with each arc. This is the key limitation in static dataflow [Den80]. It can be modeled with the recurrence in figure 5. The feedback arc begins with an initial token. This token represents a "space" on the output buffer of actor A. After A fires, and consumes that token, it cannot fire again until after B has fired. Any memory limitation on any arc in an SDF graph can be modeled as a feedback path with a fixed number of delays. To avoid unnecessarily sacrificing concurrency, enough memory should be allocated to each arc that the corresponding feedback path does not become the critical loop.

Suppose that in figure 5, actors A and B are scheduled onto different processors. In a conventional shared-memory architecture, any buffer size limitation implies handshaking at run time. In effect, the feedback path in figure 5 has to be implemented at run time, just to carry semaphores that indicate when it is safe to write to the feedforward buffer. For the ordered memory architecture, however, buffer size limitations can be statically modeled by the scheduler. They imply no additional run time overhead. In figure 5, the scheduler knows that the write from "A" and the read to "B" must alternate. Since the order of the transactions is enforced at runtime without semaphores, no additional overhead is incurred.

3.3. Manifest Iteration

In manifest iteration, the number of repetitions of a computation is known at compile time, and hence is independent of the data. It can be statically scheduled. Furthermore, it can be expressed in data flow graphs by specifying the number of tokens produced and consumed each time an actor fires. For example, actor A in figure 6 produces 10 tokens each time it fires, as indicated by the "10" adjacent to its output. Actor B consumes one token each time it fires, so it will fire ten times for every firing of actor A. In conventional programming languages, this would be expressed with a *for* loop. Nested *for* loops are easily conceived as shown in figure 6. If actors A and E fire once each, then B and D will fire ten times, and C will fire 100 times. Techniques for automatically constructing static parallel schedules for such graphs are given in [Lee87a] and [Sih90].

There is no fundamental limitation on the parallelism in figure 6 (there are no directed loops). Hence, this scheme solves the first open problem listed by Dennis in [Den75], providing the semantics of a "parallel-for" in dataflow.

180

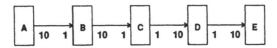

Figure 6. An SDF graph that contains nested iteration.

3.3.1. Bounded Buffers

Although there is no fundamental limitation on the parallelism in figure 6 (there are no directed loops), there may be practical limitations. In figure 7, we model a buffer of length 10 between actors B & C. Again, the tokens on the feedback path represent empty locations in the buffer. Actor B must have ten tokens on the feedback path (i.e. ten empty locations in the buffer) before it fires. Whenever actor C fires, it consumes one token from the forward path, freeing a buffer location, and indicating the free buffer location by putting a token on the feedback path. The minimum buffer size that avoids deadlock is ten.

This non-homogeneous SDF graph could be converted to a homogeneous SDF graph and the iteration period bound computed, but in this simple example the iteration period bound is easily seen by inspection. It is clear that after each firing of B, C must fire ten times before B can fire again. The ten firings can occur in parallel, so the minimum period of a periodic schedule is $R_B + R_C$, where R_X is the runtime of actor X. In other words, successive firings of B cannot occur in parallel because of the buffer space limitations. By contrast if the buffer had length 100, then ten invocations of B could fire simultaneously, assuming there are no other practical difficulties. Just as with unit length buffers, no additional synchronization overhead is required in the ordered memory architecture to support these bounded buffers.

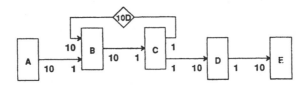

Figure 7. A modification of figure 6 to model the effect of a buffer of length ten between actors B and C.

3.3.2. Static Buffers

A second limitation on the parallelism can arise from the addressing mechanism of the buffers. Each buffer can be implemented as a FIFO queue, as done in Davis' DDM [Dav78]. Delays are correctly handled, but then access to the buffer becomes a critical section of the parallel code. FIFO queues are economically implemented as circular buffers with pointers to the read and write locations. However, parallel access to the pointers becomes a problem. If successive invocations of an actor are to fire simultaneously on a several processors, then great care must be taken to ensure the integrity of the pointers. A typical approach would be to lock the pointers while one processor has control of the FIFO queue, but this partially serializes the implementation. Furthermore, this requires that the hardware support an indivisible test-and-set operation.

In the ordered memory architecture, the FIFO implementation can be made simpler than in a general shared memory architecture, but a less expensive alternative is static buffering [Lee87b]. Static buffering is based on the observation that there is a periodicity in the buffer access that a compiler can exploit. It preserves the behavior of FIFO queues (namely it correctly handles delays and ordering of tokens), but avoids read and write pointers. Specifically, suppose that all buffers are implemented with fixed-length circular buffers, implementing FIFO queues, where each length has been predetermined to be long enough to sustain the run without causing a deadlock. Then consider an input of any actor in an SDF graph. Every N firings, where N is to be determined, the actor will get its input token(s) from the same memory location. The compiler can hard-code these memory locations into the implementation, bypassing the need for pointers to the buffer. Systematic methods for doing this, developed in [Lee87b], can be illustrated by example. Consider the graph in figure 7, which is a representation of figure 6 with the buffer between B and C assigned the length 10. A parallel implementation of this can be represented as follows:

```
FIRE A
DO ten times {
        FIRE B
        DO in parallel ten times {
                FIRE C
        }
        FIRE D
}
FIRE E
```

For each parallel firing of C, the compiler supplies a *specific* memory location for it to get its input tokens. Notice that this would not be possible if the FIFO buffer had length 11, for example, because the second time the inner DO loop is executed the memory locations accessed by C would not be the same as the first time. But with a FIFO buffer of length 10, invocations of C need not access the buffer through pointers, so there is no contention for access to the pointers. The buffer data can be supplied to all ten firings in parallel, assuming the hardware has a mechanism for doing this. In the ordered memory architecture, the ten firings cannot be initiated simultaneously, because of bus bandwidth limitations. However, they can be initiated at intervals of one shared-bus cycle. If this cycle time is small compared to the execution time of the actors, then the concurrency in the parallel for is adequately exploited.

An alternative to static buffering that also permits parallel firings of successive instances of the same actor is token matching [Arv82]. However, even the relatively low cost of some implementations of token matching [Pap88] would be hard to justify for SDF graphs, where static buffering can be used.

In figure 6 we use actors that produce more tokens than they consume, or consume more tokens than they produce. Proper design of these actors can lead to iteration constructs semantically similar to those encountered in conventional programming languages. In figure 8 we show three such actors that have proved useful in DSP applications. The first, figure 8a, simply emits the last of N tokens, where N is a parameter of the actor. The second, figure 8b, takes one input token and repeats it on the output. The third, figure 8c, takes one input token each time it fires, and emits the last N tokens that arrived. It has a self-loop used to remember the past tokens (and initialize them). This can be viewed as the state of the actor; it effectively prevents multiple simultaneous invocations of the actor.

A complete iteration model must include the ability to nest recurrences within iteration and to initialize the recurrences when the iteration begins. The SDF model can handle this. We will illustrate this with a finite impulse response (FIR) digital filter because it is a simple example. An FIR filter computes the inner product of a vector of coefficients and a vector with the last N input tokens, where N is the order of the filter. It is usually assumed to repeat forever, firing each time a new input token arrives. Consider the possible implementations using a data flow graph. A large grain approach is to define an actor with the implementation details hidden inside. This is the preferred approach in Gabriel. An alternative is a fine grain implementation with multiple adders and multipliers and a delay line. A third possibility is

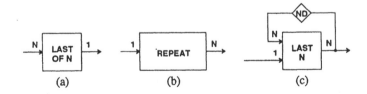

Figure 8. Three SDF actors useful for iteration.

to use iteration and a single adder and multiplier. The first and last possibilities have the advantage that the complexity of the data flow graph is independent of the order of the filter. A good compiler should be able to do as well with any of the three structures. One implementation of the last possibility is shown in figure 9. The iteration actors are drawn from figure 8. The COEFFICIENTS actor simply outputs a stream of N coefficients; it produces one coefficient each time it fires, and reverts to the beginning of the coefficient list after reaching the end. It could be implemented with a directed loop with N delays, or a number of other ways. The product of the input data and the coefficients is accumulated by the adder with a feedback loop. The output of the filter is selected by the "last of N" actor.

The FIR filter in figure 9 has the advantage of exploitable concurrency combined with a graph complexity that is independent of the order of the filter. Note, however, that there is a difficulty with the feedback loop at the adder. Recall from above that a delay is simply an initial token on the arc. If this initial token has value zero, then the first output of the FIR filter will be correct. However, after every N firings of the adder, we wish to *reset* the token on that arc to zero. This could be done with some extra actors, but a fundamental difficulty would remain. The presence of that feedback loop implies a limitation on the parallelism of the FIR filter, and that limitation would be an artifact of our implementation. Our solution is to introduce the notion of a *resetting delay*, indicated with a diamond containing an R.

3.3.3. Resetting Delays

A resetting delay is associated with a subgraph, which in figure 9 is surrounded with a dashed line. For each invocation of the subgraph, the delay token is re-initialized to zero. Furthermore, the scheduler knows that the precedence is broken when this occurs, and consequently it can schedule successive FIR output computations simultaneously on separate processors.

The resetting delay can be used in any SDF graph where we have nested iterations where the inner iterations involve recurrences that must be initialized. In other words, anything of the form:

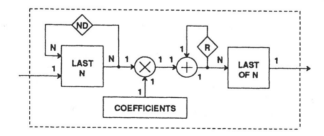

Figure 9. An FIR filter implemented using a single multiplier and adder.

```
DO some number of times {
        Initialize X
        DO some number of times {
                new X = f(X)
        }
}
```

The implementation of a resetting delay is simple and general. For the purposes of implementation, the scheduler first treats the delay as if it were an actor that consumes one token and produces one token each time it fires. Recall that in practice no runtime operation is required to implement a delay, so there actually is no such actor. However, by inserting this mythical actor, the scheduler can determine how many times it would fire (if it did exist) for each firing of the associated subgraph. The method for doing this is given in [Lee87a], and consists of solving a simple system of equations. For each resetting delay, the scheduler obtains a number N of invocations between resets; this number is used to break the precedence of the arc for every N^{th} token and to insert object code that re-initializes the delay value. The method works even if the subgraph is not invoked as a unit, and even if it is scattered among the available processors. It is particularly simple when in-line code is generated. However, when the iteration is implemented by the compiler using loops, then a small amount of run-time overhead may have to associated with some delays in order to count invocations.

So far we have shown that neither manifest iteration nor recurrences present a fundamental problem for the SDF model. Resetting delays can be used to initialize recurrences within nested iterations. Hence corresponding programming constructs can be efficiently and automatically implemented on an ordered memory architecture. Conditionals are a bit more problematic.

3.4. Conditional Assignment

Conditionals in dataflow graphs are harder to describe and schedule statically. One attractive solution is a mixed-mode programming environment, where the programmer can use dataflow at the highest level and conventional languages such as C at a lower level. Gabriel is precisely such an environment. Conditionals would be expressed in the conventional language. This is only a partial solution, however, because conditionals would be restricted to lie entirely within one large grain actor, and concurrency within such actors is difficult to exploit. If the complexity of the operations that are performed conditionally is high, then this approach is not adequate. Furthermore, conditionals within an actor usually imply a non-deterministic execution time of the actor. If the variability of the possible execution times is high, the performance of the ordered memory architecture will suffer.

A simple alternative that is sometimes suitable is to replace *conditional evaluation* with *conditional assignment*. The functional expression

$$y \leftarrow \text{if } (c) \text{ then } f(x) \text{ else } g(x)$$

can be implemented as shown in figure 10. The MUX actor consumes a token on each of the T, F, and control inputs and copies either the T or F token to the output. Hence, both $f(x)$ and $g(x)$ will be computed and only one of the results will be used. When these functions are simple, this approach is efficient; indeed it is commonly used in deeply pipelined processors to avoid conditional branches. For hard-real-time applications, it is also efficient when *one of the two* subgraphs is simple. Otherwise, however, the cost of evaluating both subgraphs may be excessive, so alternative techniques are required.

4. A NOTE ON QUASI-STATIC SCHEDULING

The domain of applications of the ordered memory architecture is constrained by the need to statically order shared memory accesses. An automatic parallelizing compiler has been written to work with SDF graphs where the actor execution times are reasonably predictable. Since the SDF model supports recurrences, manifest iteration, and conditional assignment, it is not as limited as might first appear. Nonetheless, it is worth attempting to weaken the constraints of the SDF model in order to encompass more applications. At Berkeley we have been developing *quasi-static* scheduling strategies that may solve some of these problems [Ha89]. The basic principle is that dynamic control is used only where absolutely necessary. For instance, with an if-then-else, control is dynamically transferred to one of two statically scheduled subgraphs. Similarly, for a data-dependent iteration (such as a do-while), a static schedule for each cycle of the iteration is dynamically repeated. The challenge, of course, is to develop strategies for constructing the static schedules for the subgraphs. Furthermore, these techniques imply changes to the ordered memory architecture. The MOMA controller can no longer simply passively step through a static list of processors to which it must grant the bus. Instead, it has to follow the control path of the

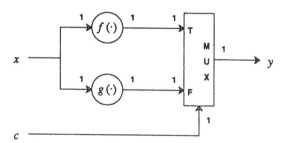

Figure 10. A dataflow graph with conditional assignment. Both $f(\cdot)$ and $g(\cdot)$ are evaluated, and only one of the two outputs is selected.

distributed executing program. The challenge is to accomplish this without increasing the complexity of the controller so much that the advantages of ordered memory accesses evaporate. This is an active and promising line of inquiry.

5. CONCLUSIONS

It is well known that data-independent dataflow graphs can be scheduled statically, obviating the need for additional runtime hardware to control the execution. We have illustrated low-cost parallel architectures that take advantage of this, and have shown that shared data structures can be also be supported efficiently. A software simulation of an ordered memory architecture has been built along with a compiler that fully automates the mapping. The compiler begins with a large grain dataflow graph that conforms with the synchronous dataflow model of computation. This SDF model, although limited, can support recurrences, manifest iteration, and conditional assignment. The execution times of actors can vary slightly without seriously affecting the implementation, but wide variations can have considerable adverse impact on execution speed. For applications with little decision making, such as signal processing and some scientific computing, this approach appears attractive. To broaden the application base, quasi-static scheduling may provide a solution by introducing dynamic control only where absolutely necessary [Ha89].

REFERENCES

[Arv82]
Arvind and K. P. Gostelow, "The U-Interpreter", *Computer* 15(2), February 1982.

[Arv87]
Arvind, R. S. Nikhil, and K. K. Pingali, "I-structures: Data Structures for Parallel Computing", Computation Structures Group Memo 269, MIT, February 1987 (revised March 1989). Also to appear in *ACM Transactions on Programming Languages and Systems*.

[Arv89]
Arvind, private communication, 1989.

[Bie89]
J. C. Bier and E. A. Lee, "Frigg: A Simulation Environment for Multiple-Processor DSP System Development", *Proceedings of the International Conference on Computer Design*, Boston, October 2-4, 1989.

[Bie90]
J. Bier and E. A. Lee, "A Class of Multiprocessor Architectures for Real-Time DSP", *Proc. of the Int. Conf. on Circuits and Systems*, New Orleans, May, 1990.

[Bur81]
F. W. Burton and M. R. Sleep, "Executing Functional Programs on A Virtual

Tree of Processors", *Proc. ACM Conf. Functional Programming Lang. Comput. Arch.*, pp. 187-194, 1981.

[Cha84]

M. Chase, "A Pipelined Data Flow Architecture for Signal Processing: the NEC uPD7281", *VLSI Signal Processing*, IEEE Press, New York (1984)

[Chu80]

W. W. Chu, L. J. Holloway, L. M.-T. Lan, and K. Efe, "Task Allocation in Distributed Data Processing", *IEEE Computer*, pp. 57-69, November, 1980.

[Chu87]

W. W. Chu and L. M.-T. Lan, "Task Allocation and Precedence Relations for Distributed Real-Time Systems", *IEEE Trans. on Computers*, C-36(6), pp. 667-679, June 1987.

[Coh85]

G. Cohen, D. Dubois, J. P. Quadrat, and M. Viot, "A Linear-System-Theoretic View of Discrete-Event Processes and its Use for Performance Evaluation in Manufacturing", *IEEE Trans. on Automatic Control*, AC-30, 1985, pp. 210-220.

[Cor79]

M. Cornish, D. W. Hogan, and J. C. Jensen, "The Texas Instruments Distributed Data Processor", *Proc. Louisiana Computer Exposition*, Lafayette, La., pp. 189-193, March 1979.

[Dav78]

A. L. Davis, "The Architecture and System Method of DDM1: A Recursively Structured Data Driven Machine", *Proc. Fifth Ann. Symp. Computer Architecture*, April, 1978, pp. 210-215.

[Den75]

J.B. Dennis, "First Version Data Flow Procedure Language", Technical Memo MAC TM61, May, 1975, MIT Laboratory for Computer Science.

[Den80]

J. B. Dennis, "Data Flow Supercomputers" *Computer*, 13 (11), November 1980.

[Den88]

J. B. Dennis and G. R. Gao "An Efficient Pipelined Dataflow Processor Architecture" To appear in *Proceedings of the IEEE*, also in the *Proc. ACM SIGARCH Conf. on Supercomputing*, Florida, Nov., 1988.

[Dri86]

J. Driscoll, N. Sarnak, D. Sleator, and R. Tarjan, "Making Data Structures Persistent", in *Proc. of the 18th Ann. ACM Symp. on Theory of Computing*, Berkeley, CA, pp.109-121, May 1986.

[Efe82]

K. Efe, "Heuristic Models of Task Assignment Scheduling in Distributed Systems", *IEEE Computer*, pp. 50-56, June, 1982.

[Fis84]

J. A. Fisher, "The VLIW Machine: A Multiprocessor for Compiling Scientific Code", *Computer*, July, 1984, 17(7).

[Gao88]

G. R. Gao, R. Tio, and H. H. J. Hum, "Design of an Efficient Dataflow

Architecture without Data Flow", *Proc. Int. Conf. on Fifth Generation Computer Systems*, 1988.

[Gau86]
J. L. Gaudiot, "Structure Handling in Data-Flow Systems", *IEEE Trans. on Computers*, C-35(6), June 1986.

[Gau87]
J. L. Gaudiot, "Data-Driven Multicomputers in Digital Signal Processing", *IEEE Proceedings*, Vol. 75, No. 9, pp. 1220-1234, September, 1987.

[Gra87]
M. Granski, I. Koren, and G.M. Silberman, "The Effect of Operation Scheduling on the Performance of a Data Flow Computer", *IEEE Trans. on Computers*, C-36(9), September, 1987.

[Ha89]
S. Ha and E. A. Lee, "Compile-time Scheduling and Assignment of Dataflow Program Graphs with Data-Dependent Iteration", Memorandum no. UCB/ERL M89/57, Electronics Research Lab, U. C. Berkeley, Berkeley, CA 94720.

[Hoa78]
C. A. R. Hoare, "Communicating Sequential Processes", *Communications of the ACM*, August 1978, 21(8)

[Hu61]
T. C. Hu, "Parallel Sequencing and Assembly Line Problems", *Operations Research*, 9(6), pp. 841-848, 1961.

[Hud86]
P. Hudak, "A Semantic Model of Reference Counting and its Abstraction", in *Proc. of the 1986 ACM Conf. on Lisp and Functional Programming*, MIT, Cambridge, MA, pp. 351-363, August, 1986.

[Ian88]
R. A. Iannucci, "A Dataflow/von Neumann Hybrid Architecture", Technical Report TR-418, MIT Lab. for Computer Science, 545 Technology Square, Cambridge, MA 02139, May 1988.

[Iqb86]
M. A. Iqbal, J. H. Saltz, and S. H. Bokhari, "A Comparative Analysis of Static and Dynamic Load Balancing Strategies", *Int. Conf. on Parallel Processing*, pp. 1040-1045, 1986.

[Jag86]
H. V. Jagadish, R. G. Mathews, T. Kailath, and J. A. Newkirk, "A Study of Pipelining in Computing Arrays", *IEEE Trans. on Computers*, C-35(5), May 1986.

[Kel84]
R. M. Keller, F. C. H. Lin, and J. Tanaka, "Rediflow Multiprocessing", *Proc. IEEE COMPCON*, pp. 410-417, February, 1984.

[Kim88]
S.J. Kim and J.C. Browne, "A General Approach to Mapping of Parallel Computations upon Multiprocessor Architectures", *Proceedings 1988 International Conference on Parallel Processing*, August, 1988, pp. 1-8.

[Kun87]
J. Kunkel, "Parallelism in COSSAP", *Internal Memorandum*, Aachen

University of Technology, Fed. Rep. of Germany, 1987.

[Kun88]

S. Y. Kung, *VLSI Array Processors*, Prentice-Hall, Englewood Cliffs, NJ (1988).

[Lee86]

E. A. Lee, "A Coupled Hardware and Software Architecture for Programmable Digital Signal Processors", *Memorandum No. UCB/ERL M86/54*, EECS Dept., UC Berkeley (PhD Dissertation), 1986.

[Lee87a]

E. A. Lee and D. G. Messerschmitt, "Static Scheduling of Synchronous Data Flow Graph for Digital Signal Processing", *IEEE Trans. on Computers*, January, 1987.

[Lee87b]

E. A. Lee and D. G. Messerschmitt, "Synchronous Data Flow", *IEEE Proceedings*, September, 1987.

[Lee88]

E. A. Lee, "Programmable DSP Architectures, Part I", *ASSP Magazine*, October, 1988.

[Lee89a]

E. A. Lee, "Programmable DSP Architectures, Part II", *ASSP Magazine*, January, 1989.

[Lee89b]

E. A. Lee, W.-H. Ho, E. Goei, J. Bier, and S. Bhattacharyya, "Gabriel: A Design Environment for DSP", *IEEE Trans. on ASSP*, November, 1989.

[Lu86]

H. Lu and M. J. Carey, "Load-Balanced Task Allocation in Locally Distributed Computer Systems", *Int. Conf. on Parallel Processing*, pp. 1037-1039, 1986.

[Ma82]

P. R. Ma, E. Y. S. Lee and M. Tsuchiya, "A Task Allocation Model for Distributed Computing Systems", *IEEE Trans. on Computers*, Vol. C-31, No. 1, pp. 41-47, January, 1982.

[Muh87]

H. Muhlenbein, M. Gorges-Schleuter, and O. Kramer, "New Solutions to the Mapping Problem of Parallel Systems: The Evolution Approach", *Parallel Computing*, 4, pp. 269-279, 1987.

[Nik89]

R. S. Nikhil and Arvind, "Can Dataflow Subsume von Neumann Computing?", *Proc. of the 16th Annual Int. Symp. on Computer Architecture*, Jerusalem, Isreal, May 28 - June 1, 1989.

[Pap88]

G. M. Papadopoulos, *Implementation of a General Purpose Dataflow Multiprocessor*, Dept. of Electrical Engineering and Computer Science, MIT, PhD Thesis, August, 1988.

[Pet77]

J. L. Peterson, *Petri Net Theory and the Modeling of Systems*, Prentice-Hall Inc., Englewood Cliffs, NJ, 1981.

[Pla76]

A. Plas, *et. al.*, "LAU System Architecture: A Parallel Data-driven Processor Based on Single Assignment", *Proc. 1976 Int. Conf. Parallel Processing*, pp. 293-302.

[Ren81]

M. Renfors and Y. Neuvo, "The Maximum Sampling Rate of Digital Filters Under Hardware Speed Constraints", *IEEE Trans. on Circuits and Systems*, CAS-28(3), March 1981.

[Sih90]

G. Sih and E. A. Lee, "Scheduling to Account for Interprocessor Communication within Interconnection-Constrained Processor Networks", *Proc. Int. Conf. on Parallel Processing*, August, 1990.

[Wat82]

I. Watson and J. Gurd, "A Practical Data Flow Computer", *Computer* **15** (2), February 1982.

[Zis87]

M. A. Zissman and G. C. O'Leary, "A Block Diagram Compiler for a Digital Signal Processing MIMD Computer", *IEEE Int. Conf. on ASSP*, pp. 1867-1870, 1987.

8

DESIGN OF ASYNCHRONOUS PARALLEL ARCHITECTURES

Teresa H.-Y. Meng
Department of Electrical Engineering
Stanford University

One major difficulty in implementing high performance parallel architectures resides in global interdependencies such as arise from clock and data synchronization. To effectively alleviate the design difficulties in global synchronization, we need to establish a modular and flexible design framework, in which the path connecting system architectures to physical implementations can be formally defined and the distance between them be reduced. Asynchronous design, which does not require a global clocking signal, provides a modular design approach to implementing parallel architectures without compromising the global performance. Asynchronous design can support fast-prototyping of parallel processing systems with a minimal amount of design effort, by providing an interconnect strategy which allows the overall system performance to be improved by individual optimization of computation modules. This allows the equivalent of the *object oriented* software methodology to be applied to hardware design. In this chapter, we will describe the fundamentals of asynchronous circuits and demonstrate the ease with which asynchronous parallel architectures can be realized through automated synthesis tools.

This research is sponsored in part by the Office of Naval Research under the ONR Young Investigator program with contract number N00014.

1. INTERCONNECTION OF SYSTEM MODULES

As we develop VLSI systems with ever faster digital circuits, the distribution and synchronization of a global clock becomes a bottleneck to system throughput. It has already become clear that system clock speeds are starting to significantly lag behind logic speeds of the underlying technology. While gate delays are well below 1 nanosecond in advanced CMOS technology, clock rates of more than 100 Mhz are difficult to attain. In order to achieve high performance, extensive design and simulation effort is required. This problem will become worse in the future as the complexity of our chips and systems, and the speed of logic, increase further.

An alternative approach is to use asynchronous design, which eliminates the need for a global clock and circumvents problems associated with global clock synchronization. At the chip design level, asynchronous interfaces can be used between modules within a chip. Since there are no global timing considerations, the design time spent on layout and circuit timing simulation can be greatly reduced. At the board design level, systems built using this approach can be easily extended without problems in global synchronization. This is particularly important for signal processing applications using parallel pipelined architectures, where computation hardware can be extended by pipelining without any degradation in overall system throughput.

In the past, the circuit and delay time overhead associated with asynchronous design was sufficiently high so that asynchronous circuits have not been widely used. However, many examples can be found now, even in critical performance areas. The VME bus interface is fully asynchronous, as a system clock is not used in validating data transfers (although the clock may be propagated over the bus). At a lower level standard state-of-the-art microprocessors, such as the 68000 series, use an asynchronous interface with a four-phase handshake for memory and I/O transfers.

One strategy is therefore as follows: a set of modules will be defined that have an asynchronous interface at their boundary. The only constraint on the modules is that they have a completion signal, which indicates when the module has completed its processing. A set of deterministic algorithms for determining the interface circuits between the modules have been developed. [1]. Based on this approach, an arbitrarily complex system with a structural description can be designed without having to be concerned with the underlying timing considerations.

1.1. Fully Asynchronous Design

In the past, the use of asynchronous circuitry is more prevalent at the higher levels of the system design hierarchy, such as the VME bus. The reason for this is that at the higher levels the complexity of interconnect is simple enough that a proper design can be made without sophisticated synthesis tools. At the lower levels of design, concerns about hazard and race conditions embedded in asynchronous logic and the inherent metastability phenomenon in arbiters are often mentioned. Another major factor is that logic has been slow relative to typical clock speeds, in which case the handshake circuit overhead is substantial.

In spite of these concerns, significant theoretical work has been done on asynchronous logic design. Asynchronous logic networks with bounded delay elements have been investigated, often referred to as the Huffman model [2]. Hazard-free logic networks can be designed by proper assumptions on gate delays and by restricting changes of inputs to a single variable at a time [3]. However, these classical design procedures for asynchronous logic are error-prone and require exponentially complex enumeration. Furthermore, the restriction on input signals puts a severe limitation on the kind of asynchronous circuits that can be realized. Research on asynchronous circuits with unbounded gate delays have used data detectors and spacers or coding schemes [4]. Again, these methods involve precaution in the design phase to encode data lines and needs at least twice the hardware to compute coded data.

One approach to eliminating some of these problems is based on the synthesis of asynchronous self-timed circuits from a high-level specification [1]. Self-timed circuits differ from the early asynchronous circuits in that they can be designed to be *speed-independent*; i.e. their behavior does not depend on any relative delays among physical gates [5,6,4]. An asynchronous self-timed circuit is called *delay-insensitive* if, besides being speed-independent, its behavior does not depend on the wire delays within a module (this module can be as large as a system or as small as a gate). As shown by Martin in [7], truly delay-independent circuits are of only limited use. Most of the asynchronous circuits discussed in this chapter are speed-independent, but not insensitive to wire delays (although they are insensitive to the wire delays *between* modules). Since asynchronous speed-independent circuits guarantee correct timing by synthesis, timing simulations are required only to predict performance but not to verify correctness.

We are primarily concerned with designing asynchronous parallel architectures of desired properties from a system point of view, as opposed to confining our attention to certain circuit modules. Therefore we start the discussion with an overview of the proposed design methodology; we call it the *fully asynchronous design approach*. It uses extra handshaking leads to avoid the distribution of a clock signal, and data synchronization is accomplished by a four-phase handshake protocol. We will also introduce a logic family which supports this protocol.

There are basically two modules in a parallel system: computation modules for operation calculation and interconnection modules for data transfers (shown in figure 1). Computation modules, which includes combinational logic such as shifters and multipliers and memory elements such as RAMs and ROMs, perform processor operations. Interconnection modules, which include data transfer circuits such as pipeline registers and multiplexers (indicated in figure 1 by blocks filled with hashed lines), handle transfers of both control and data signals. A computation module generates a completion signal to indicate valid output and to request a data transfer. An interconnection module operates on these completion and request signals and insures that data are correctly transferred regardless of the relative delays among these handshake signals.

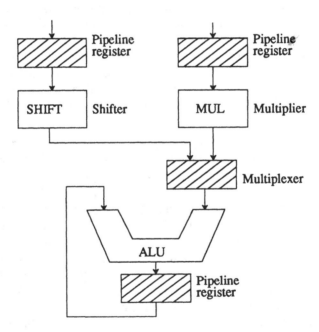

Figure 1. Using the block diagram design approach, a digital design can be specified by decoupled computation modules and interconnection modules.

The interconnection protocol used can be the standard four-phase handshake [8] or the more efficient two-phase handshake. The four-phase handshake protocol [9] allows the use of only *level* sensitive logic if interconnection modules are properly designed. The two-phase handshake protocol uses a signal *edge*, rising or falling, to indicate a data transfer. While this can be more efficient in timing overhead, to sense a signal edge will require the use of *edge* sensitive logic and thus may complicate the circuit design. In this chapter, we will use the four-phase handshake as the interconnection protocol.

1.2. Computation Modules

One type of computation modules is combinational logic for memoryless function evaluation, and the other has memory elements for program control and data accessing. It has been shown that two binary handshake signals (*request* and *complete* signals in our case) are necessary and sufficient to realize general asynchronous networks with unbounded gate delays [8]. For efficient hardware implementation, any combinational operation can be conveniently combined into one computation module, given that request and completion signals are generated along the datapath. Memory elements can be handled in the same way, as long as the memory access mechanism is initiated by a request signal and ended with a completion signal.

Our experimental implementations employ a logic family, differential cascode voltage switch logic (DCVSL) [10,11], to generate completion signals for combinational logic in a general way. A DCVSL computation module is shown in figure 2. where the *request* signal can be viewed, for the moment, as the completion signal from the preceding module. When *request* is low, the two output data lines would be pulled up by the p-MOS transistors and the *complete* signal will be low. When *request* goes high, which indicates that the computation of the preceding module has been completed and the differential input lines (both control and data) are stable and valid for evaluation, the two p-MOS transistors would be cut-off and the input lines will be evaluated by the NMOS tree. The NMOS logic tree is designed such that only one of the output data lines (two lines per data bit) will be pulled down by the NMOS tree once and only once, since there is no active pull-up device; then the *complete* signal will be set high by the NAND gate. Feedback transistors can be used at the output data lines to make DCVSL static.

Differential inputs and outputs are necessary for completion signal generation. Existing logic minimization algorithms [12] can be used to design differential NMOS logic trees, with a result that the hardware overhead is minimal compared with dynamic circuits. DCVSL has been found to offer a performance advantage compared with primitive NAND/NOR logic families, since NMOS logic trees are capable of computing complex Boolean

196

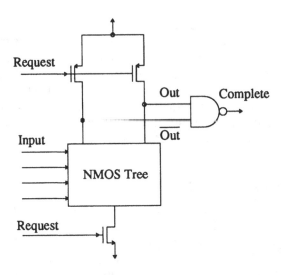

Figure 2. A schematic diagram of DCVSL for completion signal generation. The completion signal *Complete* goes high when output data lines become stable and stays low when *Request* is precharging the NMOS tree.

functions within a single gate delay [13]. Several DCVSL computation modules commonly used in signal processing, such as arithmetic-logic-units, shifters, and multipliers, have been designed and demonstrated a speed comparable to their synchronous counterparts [11,14]. The routing of differential input and output lines often requires on the order of a 40% active area penalty.

1.3. Interconnection Modules

An interconnection module operates on the completion signals generated by computation modules so that data are properly transferred. Since requests for communication may be issued at any time, an interconnection circuit is ideally designed such that no erroneous behavior will occur under all possible variations in module time and gate delays. The circuit behavior is thus independent of the relative delays among circuit elements. This kind of the circuits are usually called speed-independent or self-timed circuits.

As an example, the simplest interconnection circuit is a pipeline handshake circuit. Figure 3 shows two computation modules labeled A and B connected in a pipeline. Because of the pipeline handshake circuit, block A can process the next data sample while block B processes the current sample. Since block A might take longer to finish than block B or *vice versa*, an *acknowledge* (A_{in}) signal is necessary to indicate when block B has completed its task and is ready for the next data sample. The handshake circuit must prevent *run-away* conditions (samples being overwritten at the input to block B if block B has a long computation latency) or *continual feeding* (samples being computed more than once by block B if block A has a long latency). Other types of asynchronous handshake circuits display similar requirements.

A procedure for interconnection circuit synthesis that meets these requirements is constructed by modeling computation modules as adding uncontrolled transmission delays between the terminals of interconnection circuits. The procedure we developed allows us to optimize the design of an interconnection circuit as we perform the synthesis. Starting from a high-level specification, we describe a deterministic synthesis algorithm with no heuristics. The logic synthesized is guaranteed to be hazard-free with the fastest operations (maximum concurrency) among all the circuits that behave according to the same specification. Since the output of the algorithm is in the form of Boolean expressions, hardware complexity of the real circuit implementation can be minimized using standard Boolean minimization algorithms. During the synthesis process, designers can be interactively advised of the performance consequence of their specifications.

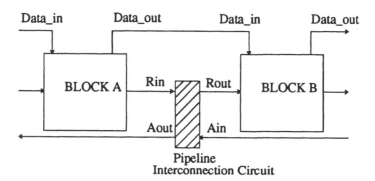

Pipeline
Interconnection Circuit

Figure 3. A simple example of an interconnection circuit: a pipeline interconnection circuit that controls data transfers between computation modules A and B.

2. INTERCONNECTION CIRCUIT SYNTHESIS

Recently asynchronous circuit synthesis has become an active research area [15,16]. Trace theory and Petri nets were both proposed for synthesis of speed-independent circuits from a behavioral description that describes circuit's trace structures [17,18]. But two consequences of these approaches are that the resulting circuits tend to be complex and that the degree of concurrency realized in the synthesized circuits cannot be directly addressed. The same problem arises in defining speed-independent modules that can be compiled together to form specific circuit functions.

Synthesis of speed-independent interconnection circuits using signal transition graphs [15] appears to be a promising approach. In a signal transition graph, the rising transition of a signal x is denoted as x^+ and a falling transition as x^-. An arc in a signal transition graph represents a causal relation: for example, $x^+ \rightarrow y^+$ means that x must have gone high before y can go high. We say that x^+ *enables* y^+ and y^+ is *enabled* by x^+. Given a signal transition graph, the synthesis process modifies the graph to satisfy a number of syntactic requirements and derives a logic implementation from the modified graph. The signal transition graph provides a clear view of potential circuit hazards and admits rules to preclude these hazards. Nevertheless, the initial stage of constructing a signal transition graph from a desired control operation is non-trivial.

2.1. Interconnection Circuit Specifications

Experience has shown that describing circuit behavior, even with the abstraction of signal transition graphs, is a tedious task and often gives suboptimum solutions because of suboptimum specifications. We need a higher level description than signal transition graphs. We therefore describe an alternative high-level specification, the *guarded command* [19].

The basic construct of a guarded command is a statement list prefixed by a Boolean expression [*precondition* → *statements*]: Only when the precondition is true, will the statement list be eligible for execution. A list of legal guarded command constructs are given in table 1. A guarded command is formed by combining basic constructs in a language syntax formally defined in [19]. The precondition specified in a guarded command is the minimum information on circuit behavior necessary for circuit synthesis.

The simplest interconnection circuit is a handshake circuit implementing the pipelined architecture shown in figure 3. The four-phase handshake protocol dictates that the sequence of signal transitions on the right hand side of the pipeline handshake circuit is always the iterative $R_{out}^+ \rightarrow A_{in}^+ \rightarrow R_{out}^- \rightarrow A_{in}^-$ and on the left hand side $R_{in}^+ \rightarrow A_{out}^+ \rightarrow R_{in}^- \rightarrow A_{out}^-$. In synthesizing the Boolean functions for A_{out} and R_{out}, if too weak a condition were used to

Deterministic Guarded Commands	
Notation	Interpretation
Basic guarded command: $[C \rightarrow S]$	C is a *pre-condition* and S is a list of statements to be executed if C is true.
AND command: $[C_1 \cap C_2 \cap \cdots \cap C_n \rightarrow S]$	C_i is a pre-condition and S is to be executed if every C_i is true.
OR command: $[C_1 \cup C_2 \cup \cdots \cup C_n \rightarrow S]$	S is to be executed if any one of C_i is true. But for the purpose of determinism, only one of C_i can be true at a time.
Sequential Construct: $[C_1 \rightarrow S_1; \ C_2 \rightarrow S_2]$	C_2 can be tested only after S_1 has been executed.
Parallel Construct: $[C_1 \rightarrow S_1 \ \| \ C_2 \rightarrow S_2]$	The two clauses $C_1 \rightarrow S_1$ and $C_2 \rightarrow S_2$ can be processed concurrently.
Alternative Construct: $[C_1 \rightarrow S_1 | C_2 \rightarrow S_2]$	S_i is executed if C_i is true, but only one of the C_i can be true at a time.
Repetitive construct: $*[C \rightarrow S]$	The clause $[C \rightarrow S]$ is to be repeatedly executed.

Table 1. Legal constructs of deterministic guarded commands.

raise A_{out} or R_{out}, run-away conditions or continual feeding might occur (this problem was addressed in the design in [20]). If too strong a condition were used to ensure proper operation, the handshake circuit may incur an unnecessary long delay in response to a request and suffer a low percentage of hardware utilization (this happens in [21]). The design of a reliable and fast handshake circuit is therefore not necessarily straightforward.

The minimal requirement for a pipeline handshake circuit is that signal transitions follow the four-phase handshake sequences. To specify these assumed sequences for every possible connection in a signal transition graph shifts attention from circuit behavior, our primary concern, to internal details of signal transitions. More importantly, it makes it difficult to determine an optimum specification. Therefore we describe a circuit property called *semi-modularity* to guide the signal transition behavior from a guarded command specification.

2.2. Weakest Semi-Modularity Conditions

Guarded commands specify the necessary conditions for the correct behavior of an interconnection circuit. To bridge the gap between this behavioral specification and its circuit implementation, we need only strengthen the guarded preconditions until they are also sufficient to form a *semi—modular* circuit. This can be defined as being when all signal transitions in that circuit have the property that once the transition is enabled, only the firing of the transition can disable it [8]. By disabling a transition we mean that the enabling condition does not hold any more. For example, if x^+ enables y^+, semi-modularity requires that only after y has actually gone high (y^+), can x go low (x^-). Semi-modularity is sometimes confused with the *persistence* condition [15] often sited in Petri net analysis. Since a signal transition graph is an interpreted Petri net, the persistence property defined on uninterpreted Petri nets cannot describe the *disabling* condition. Hence semi-modularity is a stronger condition than the persistence property defined in Petri nets. It is this stronger condition that differentiates an hazard-free circuit *implementation* from speed-independent circuit *behavior*.

We have developed a polynomial-time algorithm to construct a semi-modular signal transition graph with the weakest conditions (insuring fastest operation) from a guarded command specification. Examples will be shown in the next subsection.

The second step in the synthesis procedure is to convert this semi-modular graph to a state diagram. It can be shown that the semi-modular signal transition graph derived from a guarded command guarantees the existence of a consistent state assignment for every state. According to the state diagram, standard Boolean minimization can be used to derive a Boolean expression for each output signal. The resulting circuit is guaranteed to be deadlock-free, hazard-free, and allows maximum possible concurrency from its signal transition specification.

2.3. Synthesis Circuit Examples

A reasonable specification of the pipeline interconnection circuit shown in figure 3 can be given as

$$* [R_{in}^+ \rightarrow R_{out}^+], \tag{1}$$

meaning that if the input sample is ready (R_{in}^+), then start computation (R_{out}^+). The requirement that the output register must be empty (A_{in}^-) before a new datum can be accepted is reflected in the four-phase handshake transitions $A_{in}^- \rightarrow R_{out}^+$.

The circuit synthesized from the specification of (1) is shown in figure 4, in which a two-input C-element implements the Boolean function $C = AB + BC' + AC'$, where A and B are the input signals, C' is the previous output signal, and C is the current output signal [8]. The C-element has the property that the output signal will change to the input level only when both inputs are of the same level; otherwise the output stays unchanged.

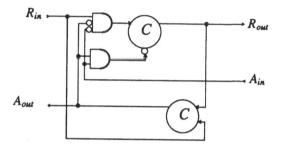

Figure 4. Logic implementation of the pipeline interconnection circuit specified by $*[R_{in}^+ \to R_{out}^+]$. Of all the circuits that behave correctly according to the specification, this circuit has the shortest response time.

The efficiency of hardware utilization resulted from the use of an interconnection circuit can be derived from its semi-modular signal transition graph. It can be shown that the specification of (1) allows only alternate computation blocks to compute at the same time, achieving at most a 50% hardware utilization [1]. We call this pipeline scheme the *half-handshake*.

Another possible specification of a pipeline interconnection circuit is

$$*[A_{out}^+ \to R_{out}^+]. \tag{2}$$

It can be shown that the interconnection circuit derived from this specification allows every computation block to compute at the same time [1]. We call this pipeline scheme the *full-handshake*. As an example, the signal transition graph shown in figure 5 is the semi-modular graph constructed from the specification given in (2), with it logic implementation shown in figure 6. Investigating (1) and (2), we do expect the specification $*[A_{out}^+ \to R_{out}^+]$ to give a full-handshake because the precondition is the acknowledge signal A_{out}, an indication to block A that block B has received the output datum from block A. As block B has received its datum, block A can proceed with calculating the next datum and of course block B can proceed with the present one; therefore both blocks can process data at the same time.

The difference in the two specifications of (1) and (2) demonstrates one of the advantages of using high-level specifications for circuit behavior description. With a high-level specification, we can exercise reasoning and analysis at the circuit behavioral level, rather than directly specifying signal transitions where the corresponding circuit performance is difficult to foresee. The high level of concurrency is what leads us to use the second specification in our synthesis and circuit implementation.

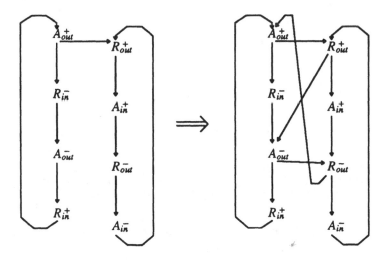

Figure 5. The weakest semi-modular transition graph constructed from the guarded command specification $* [A_{out}^+ \rightarrow R_{out}^+]$.

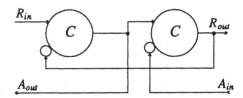

Figure 6. Logic implementation of the pipeline interconnection circuit specified by $* [A_{out}^+ \rightarrow R_{out}^+]$. This figure shows that the connection of two C-elements and two inverters is the optimal design of a *full-handshake* circuit.

2.4. Library of Interconnection Circuits

The synthesis procedure when coupled with computation modules which provide completion signals can be used to design a wide variety of circuits covering most signal processing applications. All the circuits shown in this subsection were synthesized and verified [22] to be hazard-free.

A. Sequential Interconnection Circuit:

$$* [R_{in}^+ \rightarrow R_{out}^+ ; A_{in}^+ \rightarrow A_{out}^+], \tag{3}$$

where a sequential guarded command construct is used to indicate that only after block B has completed its task can block A accept a new datum.

B. Multiple-Input Pipeline Circuit with Two Input Sources:

$$* [A_{out 1}^+ \cap A_{out 2}^+ \to R_{out}^+], \tag{4}$$

where $A_{out 1}^\downarrow$ and $A_{out 2}^\downarrow$ are ANDed in the precondition to describe that only after both of the input data are valid, can the next block start computation (R_{out}^+).

C. Multiple-Output Pipeline Circuit with Two Output Destinations:

$$* [A_{out}^+ \to R_{out 1}^+ , R_{out 2}^+], \tag{5}$$

where $R_{out 1}^+$ and $R_{out 2}^+$ are to be performed concurrently.

D. Multiplexer Pipeline Circuit with Two Input Sources

$$* [(A_{out 1}^+ \cap C_{in}^+ \cap T) \cup (A_{out 2}^+ \cap C_{in}^+ \cap \bar{T}) \to R_{out}^+], \tag{6}$$

where C_{in} is the request signal from the controller to indicate when the control signal T is valid, and T is to select one of the two input data to be transferred. $A_{out 1}^+, A_{out 2}^+$, and C_{in}^+ are specified in the precondition to insure that no operation can be performed until the control signal T and the selected datum are valid at the multiplexer input and that the full-handshake circuit is used.

E. Demultiplexer Pipeline Circuit with Two Output Destinations:

$$* [A_{out}^+ \cap C_{in}^+ \cap T \to R_{out 1}^+ \mid A_{out}^+ \cap C_{in}^+ \cap \bar{T} \to R_{out 2}^+], \tag{7}$$

where T is the control signal to select one of the output blocks. In this specification, the datum from the input block (A_{out}^+) will be transferred to block 1 $(R_{out 1}^+)$ if T is high and to block 2 $(R_{out 2}^+)$ if $\bar{T}$ is high. A_{out}^+ and C_{in}^+ are specified in the preconditions for the same reason as in a multiplexer.

The synthesized logic implementations of these interconnection circuits are shown in figure 7.

3. DESIGN CONSIDERATIONS

One important feature of asynchronous parallel architectures is that they allow the same intrinsic concurrency as data flow machines do. Asynchronous design also gives better performance than comparable synchronous design in situations for which global synchronization with a high speed clock becomes a limiting factor to the system throughput. In parallel architectures using synchronous interconnect, the throughput of the system can be limited by the global constraint of clock distribution, which in turn limits the physical size of the system, the clock rate, or both. If a parallel architecture has only localized forward-only interconnection, full pipelining in which the communication paths between chips constitute pipeline stages can

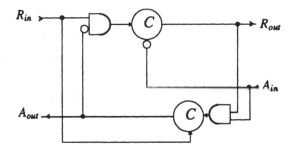

(a) Sequential Interconnection Circuit

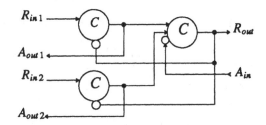

(b) Two-Input Pipeline Interconnection Circuit

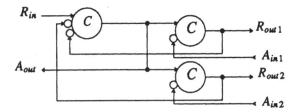

(c) Two-Output Pipeline Interconnection Circuit

Figure 7. Logic implementations of various interconnection circuits (continued).

be used and asynchronous interconnect will eliminate the global constraints on the physical size of the system. Consequently the architecture can truly achieve the throughput consistent with input/output rate limitations (such as the speed of A/D converters).

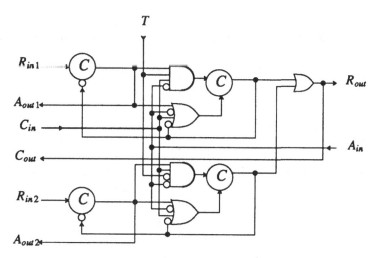

(d) Two-input Multiplexer Interconnection Circuit

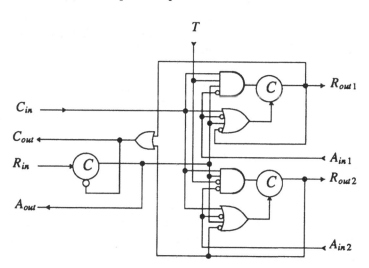

(e) Two-Output Demultiplexer Interconnection Circuit

Figure 7. Logic implementations of various interconnection circuits.

The ability to decouple the timing considerations from the design of computation modules makes the asynchronous approach particularly attractive in reducing design effort when implementing parallel architectures. In a modular design approach, the basic computation modules (parameterized macro cells) that have been designed for one system, such as multipliers, shifters, selectors, etc., can be readily used in other systems as well. Old computation modules can be replaced by new ones without modifying the original system, since asynchronous interconnect guarantees the correctness of operations without timing assumptions on individual modules. In this section, we will describe some basic properties of asynchronous modules and their design considerations with emphasis on the performance issues.

3.1. Properties of Asynchronous Modules

DCVSL computation modules use the so-called *dual–rail* [23] coded signals inside the logic, which compute both the data signal and its complement. Since both rising and falling transitions have to be completed before the completion signal can go high, if there is no carry-chain in the design, DCVSL computation modules usually give the same worst-case computation delay independent of input data patterns. This lack of data dependency to the first order is actually an advantage for real-time signal processing, since the worst case computation delay can be easily determined.

Once an asynchronous processor has been designed, there is no way to slow down the internal computation. The throughput can be controlled by issuing the request signal at a certain rate, but the computation within the processor is performed at the full speed. This property has an impact on testing: unlike synchronous circuits, we cannot slow down the circuitry to make it work with a slower *clock*.

In asynchronous design it is tempting to introduce shortcuts that improve performance at the expense of speed-independence. However, these shortcuts necessitate extensive timing simulations to verify correctness under varying conditions. To minimize design complexity we therefore generally stayed true to the speed independence design, with the result that timing simulations were required only to predict performance and not to verify correctness.

We require that an interconnection module be speed-independent; i.e., the circuit's behavior is independent of the relative gate delays among all the physical elements in the circuit. Since the synthesis procedure uses a deterministic algorithm to maximize circuit concurrency and minimize Boolean expressions, the interconnection circuits synthesized usually require much less hardware complexity than those synthesized from a collection of compiled modules [18,24]. However, the time-delay overhead incurred by the circuit delays cannot be ignored, and will be discussed in Section 4.3 as part of the system performance evaluation.

3.2. Speed-Independence

For the convenience of our discussion here, the connection of a full-handshake circuit to its pipeline register is shown in figure 8. The handshake circuit uses the rising edge of output acknowledge signal A_{out} to

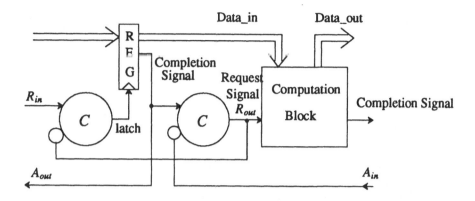

Figure 8. The connection between the pipeline register, the pipeline handshake circuit, and computation modules.

control register latching. Now we take a closer look at the latch. The register shown in figure 8 is an edge-triggered latch. The reason that an edge-triggered latch is needed instead of a level-triggered one is for fast acknowledge signal feedback. Notice that in figure 8, R_{out} (request signal) is fedback to the first C-element without waiting for the completion of the computation module connected to it. If a level triggered latch were used, the feedback signal to the first C-element will have to wait for the completion of the succeeding computation module to prevent an early precharge of its input data [25]. It can be easily verified that if a feedback signal is generated after the completion of the succeeding block, the handshake circuit would enforce that only alternate blocks can compute at the same time and that the hardware utilization is reduced to at most 50%. An edge-triggered latch does not constitute a practical problem in our design since it is strictly a local operation. The latch can be designed such that the operation is not sensitive to the slope of the triggering signal.

In order to insure the speed-independence of the latching operation, a completion detection for register latching is shown in figure 9. The logic compares the register input and output data bits. If the latching signal is high and the input and output data bits are identical, the completion signal goes high. During the reset cycle, a low on the latching signal will immediately reset the completion signal and therefore incurs only a short delay for resetting. After the data bits have been latched and stable, the completion signal from the register is fed forward as an input to the second C-element. We could match the delays of register latching and the second C-element to gain some performance improvement. However, simulation showed that any attempts at gate delay matching proved to be unreliable, and thus we chose

208

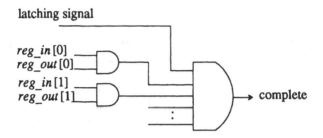

latching signal

reg_in [0]
reg_out [0]

reg_in [1]
reg_out [1]

complete

Figure 9. The completion detection mechanism for register latching. The logic compares the register input and output data bits. If the latching signal is high and the input and output data bits are identical, the completion signal goes high.

to stay true with a speed-independent design.

The data latching mechanism shown in figure 8 is gate-delay-insensitive (speed-independent) but not wire-delay-insensitive, as the wire delays of data lines may be different from that of the request signal R_{in}. If the request signal and data lines do not match in wire delays, differential pairs of data lines, or coded data lines, will have to be transmitted and the completion signal R_{in} can be generated locally at the input to the C-element.

3.3. A Self-Timed Array Multiplier

In this subsection we describe the design of a self-timed array multiplier to illustrate the performance considerations often encountered in designing DCVSL computation modules and the solutions to these problems.

3.3.1. Architecture of the Multiplier

A multiplication in most signal processing applications is usually followed by an accumulation operation, and the final addition in the multiplication can be combined with the accumulation to save the delay time of one carry-propagate add. As shown in figure 10, the multiplier core outputs two partial products to be added in the succeeding accumulate-adder, which consists of one carry-save add and one final carry-propagate add. The multiplier core is composed of three basic cells: registers, Booth encoders [26], and carry-save adders. Eight partial products are computed and added using a modified Wallace tree structure [27]. The architecture of the multiplier core shown in figure 10 consists of eight Booth encoders operating concurrently and six carry-save adders with four of them operating sequentially. The *inv* signals from encoders to carry-save adders implement the plus-one operation in a two's complement representation of a negative number. The

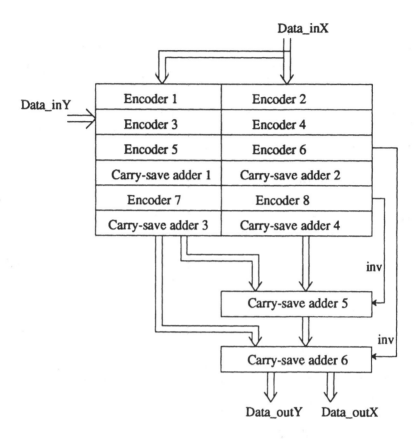

Figure 10. The architecture of the multiplier core, which consists of eight Booth encoders operating concurrently and six carry-save adders with four of them operating sequentially.

multiplier core occupies a silicon area of 2.7mm×3mm in a 1.6μm CMOS design, and a computational latency of 40 *ns* was estimated from simulation using *irsim* [28] with the 1.6μm CMOS parameters. The test results of the design will be given in the next section.

To increase the system throughput, a pipeline stage was added between the multiplier core and the accumulate-adder. We could have deep-pipelined the multiplier core as well; however, since the computational delay of the multiplier core has been reduced to only one Booth encoding plus four carry-save adds for 16-bit data, the hardware and delay-time overhead incurred in adding more pipeline stages, such as the circuitry for pipeline registers and the delay time for register latching, completion detection, and handshake operation, rules out the practicability of having one more stage of pipelining within the multiplier core.

The accumulate-adder consists of a carry-save adder at the front and a carry-propagate adder at the end to add three input operands per cycle; two of the three operands come from the output partial products of the multiplier core, with the third from a second source. The accumulate-adder occupies a silicon area of 1.1mm×1mm in a 1.6μm CMOS design. A propagation delay of 30 ns was estimated by simulation using the 1.6μm CMOS parameters and the test results will be given in the next section.

3.3.2. Circuit Design for the Multiplier Core

The goal in designing the multiplier core was to reduce the DCVSL delay by allowing maximal concurrent operation in both precharge and evaluation phases. Layout is another important design factor since a Wallace tree structure consumes more routing area than sequential adds. Data bits have to be shifted two places between each Booth encoder array to be aligned at the proper position for addition. Since each data bit is coded in a differential form, four metal lines (two for the sum and two for the carry) have to be shifted two places per array, which constitute on the order of a 40% routing area overhead as compared to similar structures in a synchronous design. The basic cells used in the multiplier core, modified from a design provided by Jacobs & Brodersen [11], will be given in this subsection and the strategies of minimizing precharge and completion generation delay overhead will be discussed.

The Carry-Save Adder

As an example of a DCVSL computation module, figure 11 shows the circuit design of a carry-save adder. Modifications on transistor sizes for better driving capability and on the PMOS transistors to precharge every node high to prevend charge-sharing were manually tailored for the specific needs of this multiplier design. As shown in figure 11, a carry-save adder takes three input bits A, B, and C and adds them up to produce two output bits $CARRY$ and SUM. Since there is no carry propagate between digits, a carry-save adder performs an addition in one gate delay.

Concurrent Precharge

Even though the five rows of sequential DCVSL gates (one for the Booth encoding and the other four for carry-save adds) evaluate data sequentially, they can be precharged concurrently. The precharge signal (R_{out}) is driven through three levels of distributed buffers to drive more than 320 PMOS precharge transistors at the same time. When the outputs of the last row of the DCVSL gates are reset to low (the inverse of a precharged high), the multiplier completion signal will go low, indicating precharge has been

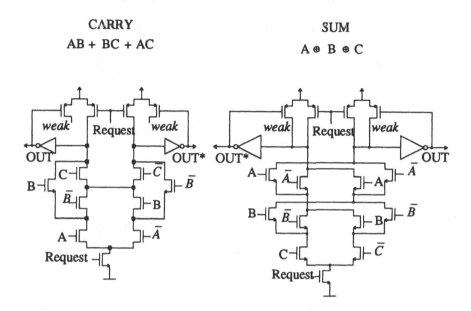

CARRY

AB + BC + AC

SUM

A ⊕ B ⊕ C

Figure 11. The circuit schematic of a DCVSL carry-save adder. Every internal node is precharged high during the precharge phase to avoid the charge sharing problem. A carry-save add consists of two separate NMOS trees, a carry tree and a sum tree.

completed. By using the concurrent precharge scheme, a 16-bit multiplier precharge operation can be finished within 6 *ns* in a 1.6μm CMOS design.

Concurrent Request

We had assumed that when the request is high, the input data to their corresponding DCVSL gates must be valid and stable. This assumption is violated by a concurrent request since DCVSL gates are connected sequentially within a pipeline stage. For example, when the request is set high, the input data to the second row of sequentially connected DCVSL gates, which are the outputs of the first row of the DCVSL gates, are definitely not valid yet. But since every output bit of DCVSL is reset to low in precharge, it is guaranteed that when the request is high at a DCVSL gate, the input bits to that gate can only go from low to high, but never from high to low. Since DCVSL consists of only NMOS logic trees, input gate voltages going from low to high will only increase the conductivity of an NMOS tree but never the reverse; therefore the one directional transition property is preserved, similar to the operation of Domino dynamic circuits.

If all the transistors along an NMOS tree are driven high, the output line will eventually go high, through an inverter at the top of the NMOS tree, indicating the completion of the evaluation for that particular tree. Valid data travel through multiple DCVSL gates sequentially, but the request signal for all the DCVSL gates are set high at the same time. After every differential pair of the last row reaches a high at one of the differential output data lines, a completion signal for the whole DCVSL block can be generated. The concurrent request scheme takes advantage of the asynchronous nature of circuit delays in order to obtain the shortest overall computational delay during the evaluation phase. Since completion signal generation can be viewed as a synchronization point, the number of unnecessary completion detections should be minimized. A completion signal is only generated at the output of the last row of a DCVSL block.

Completion Signal Generation

The completion signal for every output data pair of a DCVSL gate can be generated through an OR gate with inputs connected to the differential output data pair. As shown in figure 12, the completion signal for multi-bit data can be generated by a C-element tree which detects the completion of every data pair. In our design, two-level AND gates were used to detect the 16 completion signals at the output of a DCVSL gate, based on the assumption that precharge delay is approximately a constant for all data bits.

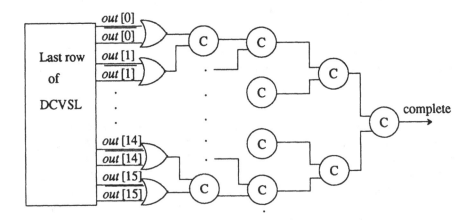

Figure 12. The completion detection for a 16-bit DCVSL block. C-elements can be replaced by simple AND gates if precharge delay is approximately constant for all data bits.

To reduce the overhead incurred by completion signal generation, we can parallel computation and completion signal generation, using the *pipelined completion detection* scheme [25]. In this scheme, level-triggered latches would have been necessary to latch both the primary and complementary data pairs. Level-triggered latches will require the acknowledge signal to be fedback only after the completion of the succeeding computation block. The overall performance gain of using pipelined completion compared to the edge-triggered latching mechanism shown in figure 8 will depend on the actual circuit functionality. In general, simple iterative computation will benefit from using the pipelined completion detection scheme, as the overhead introduced by low hardware efficiency could be more than compensated by avoiding the delay for completion signal generation.

Another obvious alternative to the completion signal generation would be to look at the NMOS tree which is expected to display the longest delay. This scheme usually works fine if there is no carry chain in the computation, as in this case each NMOS tree would exhibit the same amount of computational delay independent of input data patterns. The tradeoffs between adopting a speed-independent design and using a compromised scheme for better performance are among the most interesting issues raised for future study.

3.3.3. Circuit Design for the Accumulate-adder

The accumulate-adder is a much simpler design than the multiplier core since only one carry-save adder and one carry-propagate adder are needed. Subtraction is realized by an *inv* bit to control a simple DCVSL tree which selects either the primary data or their complements at the output of input registers and feeds the selected datum to a row of carry-save adders. The outputs of the carry-save adders are fed directly into a row of carry-propagate adders for computation of one single sum out of three input operands.

A carry-propagate adder is implemented by feeding the carry-bit of a carry-save adder to one of the three inputs of the next carry-save adder. The concurrent precharge and request plays an important role here; if a sequential precharge and request scheme had been used, one 16-bit carry-propagate adder would have incurred 16 sequential precharge delays and 16 sequential requests and completional detections, which would constitute a total delay of more than 200 ns!

The relative computational delays as exhibited by this carry-propagate adder are rather interesting. The DCVSL carry gate is designed such that the carry will go high immediately after two of the input bits (A and B) are high, regardless of the level of the third (C). If the carry from the previous DCVSL stage is taken as input C to the present DCVSL stage, two 1's on A and B will issue a carry to the next DCVSL stage without waiting for the carry from the previous stage, similar to the design of the Manchester carry chain. Consequently two 1's or two 0's from the two inputs data at the same bit position will cut the carry-chain into two independent chains, which means that the two carry chains can then compute concurrently. From the viewpoint of data-dependent computational delays, the more the concurrent 1's and 0's the faster the carry-propagate will be.

4. A VECTORIZED ADAPTIVE LATTICE FILTER

We chose to demonstrate the ease of designing asynchronous parallel architectures using the vectorized adaptive lattice filter introduced in [29]. The vectorized adaptive filter allows arbitrarily high sampling rates by exploiting unlimited parallelism within the filter algorithm. This allow us to test the highest throughput achievable through an asynchronous implementation and the effort incurred in designing such a system.

4.1. The Filter Architecture

Linear adaptive filtering is a special case of a first-order linear dynamical system. The adaptation of a single stage of a lattice filter can be described with two equations, the state-update equation

$$k(T) = a(T)k(T-1) + b(T), \tag{8}$$

and the output computation

$$y(T) = f(k(T), T), \tag{9}$$

where $a(T)$ and $b(T)$ are read-out (memoryless) functions of the input signals at time T, and $y(T)$ is a read-out function of both the state $k(T)$ and the input signals at time T. Since the computation of $y(T)$ is memoryless, its calculation can be pipeline-interleaved [30] and the computation throughput can be increased without a theoretical limit. However, the state-update represents a recursive calculation which results in an iteration bound on the system throughput [31]. In order to relax the iteration bound without changing the system's dynamical properties, (8) can be written as

$$k(T+L-1) = a(T+L-1)a(T+L-2)\cdots a(T+1)a(T)k(T-1)$$
$$+ a(T+L-1)a(T+L-2)\cdots a(T+1)b(T) + \cdots$$
$$+ a(T+L-1)b(T+L-2) + b(T+L-1) \tag{10}$$
$$= c(T,L)k(T-1) + d(T,L)$$

where $c(T,L)$ and $d(T,L)$ are memoryless functions of past input signals, but independent of the state $k(T)$. $c(T,L)$ and $d(T,L)$ can be calculated with high throughput using pipelining and parallelism, and the recursion

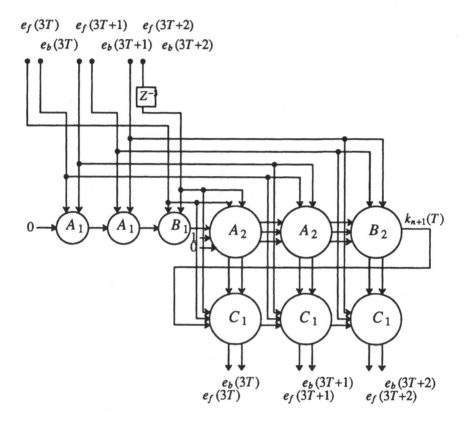

Figure 13. An LMS adaptive lattice filter stage with a vector size of three.

$k(T+L-1) = c(T,L)k(T-1) + d(T,L)$ needs only be computed every L samples. Therefore the iteration bound is increased by a factor of L. L is called the *vector size*, since a vector of L samples will be processed concurrently.

A normalized least-mean-squares (LMS) or stochastic gradient adaptation algorithm is chosen for our example because of its simplicity and the absence of numerical overflow compared to transversal recursive least-squares algorithms. A vectorized LMS lattice filter stage with $L=3$ is shown in figure 13 and the operations that each processor needs to perform are shown in figure 14. The derivation of the algorithm for normalized LMS lattice filters can be found in [24]. Processors A_1 and A_2 jointly calculate the $c(T,L)$ and $d(T,L)$ in (12), processor B_1 and B_2 calculate the state-update $k(T)$, and processor C_1 performs the output computation $y(T)$.

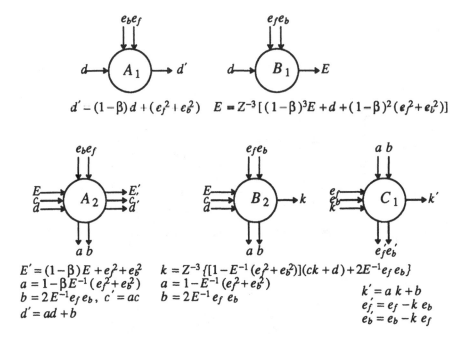

Figure 14. Operations required of each processor shown in figure 13.

For every processing cycle, a vector of L input samples are processed and a vector of L output samples are calculated. Therefore the filter sampling rate is L times faster than the processing speed. Since there is no theoretical limit to the number of samples allowed in a vector, the filter sampling rate is not constrained by the processing speed, but rather it is limited only by the I/O bandwidth (such as the speed of A/D converters). However, the higher sampling rate is achieved at the expenses of additional hardware complexity and system latency, and very high sampling rates will require multi-chip realizations. For example, at the sampling rate of 100MHz the computational requirement is approximately 3 billion operations per second per lattice stage. Since there is no global clock routing at the board level in an asynchronous design, pipelined computation hardware can be easily extended without any degradation in overall throughput as both the vector size and the number of lattice filter stages are increased.

Our goal is to design a set of chips that can be connected at the board level to achieve any sampling rate consistent with I/O limitations. This requires the partitioning of a single lattice filter stage into a set of chips which can be replicated to achieve any vector size. In this partitioning we attempted to minimize the number of different chips that had to be designed, allow

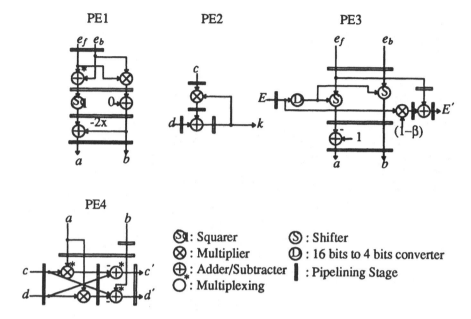

Figure 15. A chip set designed to implement the LMS adaptive lattice filter shown in figure 13. The partition was done in such a way that any required filter sampling rate can be achieved by replicating these chips at the board level without redesign. PE1 and PE4 have built-in hardware multiplexers so that the chip function can be controlled by external signals to form a reconfigurable datapath.

flexibility in the vector size (ie. amount of parallelism), and minimize the data bandwidth (number of pins) on each chip.

We found a partitioning that uses five different chips and meets these requirements. Block diagrams of four of them are shown in figure 15, and a fifth chip simply implements variable-length pipeline stages. Two of the four chips (PE1 and PE4) have built-in hardware multiplexers so that the chip function can be controlled by external signals to form a reconfigurable datapath; otherwise eight different chip designs would have been required. The same chip set can also be used to construct a lattice LMS joint process estimator. These chips can be replicated as required and interconnected at the board level to achieve any desired system throughput.

The parallel architecture of connected computation chips and pipeline stages for a vectorized LMS lattice filter with vector size $L = 3$ is shown in figure 16. Pipeline registers, indicated by rectangles in figure 16, are the most commonly required circuit elements. Since the number of pipeline stages preceding different computation chips ranges from 1 to $3L + 7$, one pipeline

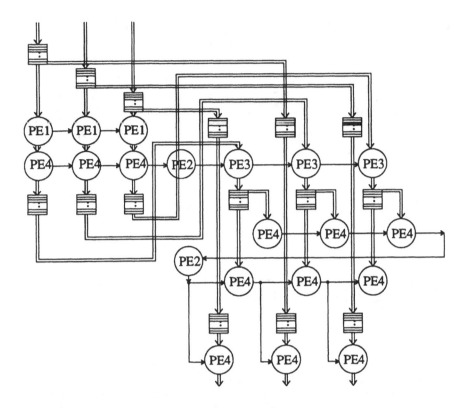

Figure 16. The connection among computation chips and pipeline registers for a vectorized LMS lattice filter stage with the vector size equal to three.

chip was designed which has a variable length and can be cascaded.

4.2. A Chip Set for Parallel Adaptive Lattice Filters

4.2.1. The Pipeline Chip

In the pipeline chip, each pipeline register is controlled by a pipeline handshake circuit and the number of pipeline stages can be specified by external signals. This chip contains two chains of pipeline stages where each chain can be programmed by external signals to act as a variable-length pipeline for up to 16 stages. The chip size is 6.4mm×7.2mm (including pads) in a 1.6 μm CMOS design. The estimated delay from pin-in to pin-out is 30 ns from simulation using the 1.6 μm CMOS parameters. This delay is considerable and due to the large capacitance on the common bus lines, each of which is connected to seven tri-state buffer diffusions. Fortunately this delay has no impact on the system throughput, as in an pipelined architecture system throughput is dominated by the pipeline stage with the longest delay, which in our case is the multiplier stage.

4.2.2. The Computation Chips

The four computation chips shown in figure 15 were designed by connecting various computation (multipliers, shifters, etc.) and interconnection (full-handshake, multiplexing, etc.) macrocells using an automatic place and route tool [14]. Here we take chip PE4 as an example as it is the most complicated one among the four.

PE4 performs three different calculations depending on how the multiplexers are configured by external signals. The multiplexers were designed such that both the feed-forward request signals and the feedback acknowledge signals are properly selected. Run-time penalty is paid for having these multiplexing operations, but three different chips would have been necessary if multiplexers were not used. Since the processing cycle time is determined by the feedback loop in PE2, the multiplexing overhead as introduced here will not degrade the overall system performance. PE4 occupies a silicon area of $7.6mm \times 7.6mm$ (including pads) with 33.2K transistors using a $1.6\mu m$ CMOS design.

4.2.3. Testing Circuitry

Scan-path registers can replace pipeline registers for testing purposes. By holding interconnection signals in a steady state, test vectors can be shifted in and out of scan-path registers within asynchronous processors as is typically done in synchronous testing, and the internal states of an asynchronous processor can be both observable and controllable.

In the pipeline chip, the output data bits at every pipeline stage is observable through external control; therefore no testing circuitry is necessary.

In the computation chips, since we designed DCVSL computation modules with registers combined in the layout to save chip area, adding extra scan-path registers into the datapath would complicate the floor plan and increase the run-time delay. Instead we connected internal completion signals to pinouts for testing purposes, because these signals are derived directly from DCVSL output data lines. If a fabrication error happened along the datapath, some completion signal would stay low as DCVSL outputs are dual-rail coded. The DCVSL macrocell in which the error resides can thus be identified. Functional errors, if any, can be caught through input test vectors.

5. PERFORMANCE EVALUATIONS

This section will give the measured performance of the chip set and describe their assembly onto circuit boards using asynchronous interconnect.

5.1. The Chip Set

The chip set was fabricated through MOSIS in 1.6μm CMOS technology and tested using the Tektronic DAS9100. The test results are summerized in table 2. The computational delay for the 16×16-bit multiplier (two output partial products) and the 16-bit accumulator (three input operands) are comparable to their synchronous counterparts. However, substantial overhead, compared to *perfect* synchronous design, is paid for the asynchronous design in the handshake delay, precharge, and completion signal generation.

Handshake delays are on the order of a couple gate delays. Delays for precharge and completion detection are usually a function of the data bit width. In our design, since we follow the strictest speed-independent rule, the operations of precharge, handshake, computation, and completion detection are processed sequentially. Even under this extreme conservative condition the processing rate can be as high as 16Mhz for one 16×16 multiplication.

The feasibility of designing asynchronous circuits with higher performance than their synchronous counterparts at the *chip level* is an issue of current debate. To answer this question, more investigation has to be made exploiting all possible parallelism in the asynchronous design regime. For example, the delay incurred by completion signal generation can be reduced [32] as discussed in the previous section. The work presented here is the first attempt to try to assess the advantages and disadvantages of using asynchronous design for parallel architectures by actually implementing these chips.

16-bit Macro	Multiplier	Accum.	Register	C-element
Computation/Set	30 ns	20 ns	4 ns	4 ns
Precharge/Reset	6 ns	6 ns	-	1 ns
Completion Sig.	8 ns	8 ns	-	-

Table 2. The test results of the computation chips using DCVSL and pipeline handshake circuits. The computation latencies are comparable to their synchronous counterparts, but the overhead paid for precharge and completion signal generation is substantial.

As will be discussed in the next subsection, the modular design approach as promised by using asynchronous components is expected to give performance advantages at the *board level*, even though an individual chip may not function at the same speed as a perfectly matched synchronous chip.

5.2. Board Level Interconnection

Systems built using a clock-free approach can be easily extended without problems in global synchronization. This is particularly important for our vectorized lattice filter application, where we want to use the same chip set to achieve a variety of sampling rates (depending on application) at the board level.

In our vectorized lattice filter application, a lower throughput, say as a result of asynchronous design or slower technology, can be compensated by a larger vector size (more parallelism) for a constant filter sampling rate. The total chip count for the computation chips increases linearly with the vector size, while the chip count for the pipeline chips increases quadratically. The number of chips of each type required for a 100MHz LMS lattice filter stage is given in table 3. As one benchmark, for a vector size of 8, 53% of the chip count is for pipelining. This overhead is due to the fact that on the order of L^2 data samples are stored within the system, which is a consequence of trading latency for throughput.

The large number of chips needed for a vectorized lattice filter was caused in part by the fact that the chip set was designed for programmable rates. In order to make the filter sampling rate parameterizable at the board level, the computation partition was constrained by the three criteria as mentioned in Section 4.1, which leads to the chip set and chip counts shown in table 3. If the design were targeted on a fixed rate, which will reduce the chip count by at least 50%, but the filter sampling rate would then be dependent on the

Cycle Time	Vector Size	Pipe. Chip	PE1	PE2	PE3	PE4	Total Counts
20ns	2	8	2	2	2	7	21
50ns	5	30	5	2	5	19	61
80ns	8	55	8	2	8	31	104
100ns	10	89	10	2	10	39	150
200ns	20	221	20	2	20	79	342

Table 3. A summary of the various numbers of chips required for a 100MHz LMS lattice filter stage using the chip set in figure 15. Over 50% of the chip count is dedicated to pipeline registers.

fabrication process. To accommodate process variations, design that exceeds specs is generally adopted, which would increase the design effort, and sometimes even the design cost.

The board level design in which the same chip set is used to achieve different levels of parallelism can be accomplished by an automated netlist generation, without worrying about clock distribution, and the desired parallelism can always be achieved with a sufficiently large vector size. Communication latencies at the board and backplane level are not of concern because they are included within pipeline stages, and are automatically compensated by the handshaking. Building different systems with varying sampling rates using the chip set and this design approach should therefore require little design effort. On the other hand, a lower hardware cost for a given sampling rate at the expense of a considerably higher design cost could be obtained in current technologies using a synchronous design methodology, or even with an asynchronous design dedicated to a given rate.

6. SUMMARY

In this chapter, the modularity which can be achieved with an asynchronous design approach was demonstrated by the presentation of an architecture and chip partitioning for an adaptive lattice filter stage with unlimited parallelism. The conservative speed-independent design strategy that we have adopted simplifies the design process at the expense of additional hardware and time-delay overhead of asynchronous interconnection circuits.

The relative performance of synchronous and asynchronous design is highly technology dependent. At the feature sizes used in our design, a synchronous design would undoubtably have higher throughput at the chip level (there will be a crossover at some smaller feature size). However, from the perspective of ease of system design and performance at the system level, an asynchronous design may be a winner.

We feel that asynchronous design is a feasible and promising approach to future parallel architecture implementation, through the design approach of separated computation and interconnection modules and an automated procedure for synthesizing interconnection circuits from a high level specification. By providing a modular design framework in which designers can prototype their systems based entirely on local properties without worrying about global variations, we believe that the efforts required in implementing complex parallel architectures can be greatly reduced. What is needed in the future is the aggressive pursuit and improvement of both synchronous and asynchronous design methodologies, and a comparison of their performance and design difficulty when applied to common applications using common technologies.

REFERENCES

1. T. H.-Y. Meng, R. W. Brodersen, and D. G. Messerschmitt, "Automatic Synthesis of Asynchronous Circuits from High Level Specifications," *IEEE Trans. on CAD*, (November 1989).

2. D. A. Huffman, "The Synthesis of Sequential Switching Circuits," *J. Franklin Institutes* **257** pp. 161-190, 275-203 (March and April 1954).

3. S. H. Unger, "Asynchronous Sequential Switching Circuits with Unrestricted Input Changes," *Trans. on Computers* C-20(12) pp. 1437-1444 (Dec. 1971).

4. S. H. Unger, *Asynchronous Sequential Switching Circuits*, Wiley-Interscience, John Wiley & Sons, Inc., New York (1969).

5. C. L. Seitz, "Self-Timed VLSI Systems," *Proc. of the Cal Tech Conference on VLSI*, (Jan. 1979).

6. D. E. Muller, "Infinite Sequences and Finite Machines," *Proc. 4th Annual IEEE Symposium on Switching Circuit Theory and Logical Design* S-**156** pp. 9-16 (Sept. 1963).

7. A. J. Martin, "The Limitations to Delay-Insensitivity in Asynchronous Circuits," *Proc. of the Sixth MIT Conference in Advanced Research in VLSI*, pp. 263-278 (May 1990).

8. R. E. Miller, *Switching Theory*, John Wiley & Sons, Inc., New York (1965).

9. C. Mead and L. Conway, *Chap. 7, Introduction to VLSI Systems*, Addison-Wesley Publishing Company (1980).

10. L. G. Heller and W. R. Griffin, "Cascode Voltage Switch Logic: A Differential CMOS Logic Family," *1984 IEEE ISSCC Digest of Technical Papers*, (Feb. 1984).

11. G. M. Jacobs and R. W. Brodersen, "Self-Timed Integrated Circuits for Digital Signal Processing Applications," *VLSI Signal Processing III*, IEEE PRESS, (November, 1988).

12. R. K. Brayton and C. McMullen, "Decomposition and Factorization of Boolean Expressions," *Proc. IEEE ICAS*, (May, 1982).

13. C. K. Erdelyi, W. R. Griffin, and R. D. Kilmoyer, "Cascode Voltage Switch Logic Design," *VLSI Design*, (October 1984).

14. T. H.-Y. Meng, R. W. Brodersen, and D. G. Messerschmitt, "Implementation of High Sampling Rate Adaptive Filters Using Asynchronous Design Techniquess," *VLSI Signal Processing III*, IEEE PRESS, (November, 1988).

15. T. A. Chu, "Synthesis of Self-Timed Control Circuits from Graphs: An Example," *Proc. IEEE 1986 ICCD*, pp. 565-571 (Oct. 1986).

16. A. J. Martin, "The Design of a Self-Timed Circuit for Distributed Mutual Exclusion," *Proc. 1985 Chapel Hill Conference on Very Large Scale Integration*, pp. 245-283 Computer Science Press, (1985).

17. J. van de Snepscheut, "Trace Theory and VLSI Design," *Lecture Notes on Computer Science 200*, Springer-Verlag, (1985).

18. D. Misunas, "Petri Nets and Speed Independent Design," *Communications of ACM* 16(8) pp. 474-481 (Aug. 1973).

19. E. W. Dijkstra, "Guarded Commands, Nondeterminacy and Formal Derivation of Programs," *Communications of the ACM* 18(8) pp. 453-457 (Aug. 1975).

20. S. Y. Kung and R. J. Gal-Ezer, "Synchronous versus Asynchronous Computation in Very Large Scale Integration Array Processors," *SPIE, Real Time Signal Processing V* 341(1982).

21. D. M. Chapiro, "Globally-Asynchronous Locally-Synchronous Systems," *Ph.D. Thesis*, Stanford University (STAN-CS-1026), (Oct. 1986).

22. D. L. Dill, *Trace Theory for Automatic Hierarchical Verification of Speed-Independent Circuits*, MIT Press (1989).

23. D. B. Armstrong, A. D. Friedman, and P. R. Manon, "Design of Asynchronous Circuits Assuming Unbounded Gate Delays," *IEEE Trans. on Computers* C-18(12)(Dec. 1969).

24. C.H. van Berkel and R. W.J.J. Saeijs, M. J. Honig, and D. G. Messerschmitt, "Adaptive Filters: Structures, Algorithms, and Applications," *Proceedings of ICCD 1988*, Kluwer Academic Publishers, (1984).

25. T. E. Williams, M. Horowitz, R. L. Alverson, and T. S. Yang, "A Self-Timed Chip for Division," *Advanced Research in VLSI, Proc of 1987 Stanford Conference*, pp. 75-96 (March 1987).

26. A. D. Booth, "A Signed Binary Multiplication Technique," *Q. J. Mech. Appl. Math.* 4(2) pp. 236-240 (1951).

27. M. Santoro and M. Horowitz, "A Pipelined Iterative Array Multiplier," *IEEE ISSCC 88 Digest of Technical Papers*, (February, 1988).

28. M. Horowitz, *IRSIM User's Manual*, Stanford University, (1988).

29. T. H.-Y. Meng and D. G. Messerschmitt, "Arbitrarily High Sampling Rate Adaptive Filters," *IEEE Trans. on ASSP* ASSP-35(4)(April 1987).

30. H.-H. Lu, E. A. Lee, and D. G. Messerschmitt, "Fast Recursive Filtering with Multiple Slow Processing Elements," *IEEE Trans. on CAS* CAS-32(11)(November, 1985).

31. M. Renfors and Y. Neuvo, "The Maximum Sampling Rate of Digital Filters Under Hardware Speed Constraints," *IEEE Trans. on CAS* CAS-28(3)(March, 1981).

32. T. E. Williams, "An Overhead-free Self-timed Division Chip," *Stanford Technical Report*, (August 1990).

9

Implementation of Multilayer Neural Networks on Parallel Programmable Digital Computers

Soheil Shams and K. Wojtek Przytula

Hughes Research Laboratories
Malibu, California 90265

ABSTRACT

Neural networks are an attractive new technology for signal processing applications, due to their adaptive, self-organizing, fault tolerant, and non-linear capabilities. An example of such an application, which is used to illustrate the results of the paper, involves a use of a multilayer perceptron network with error back-propagation learning for underwater target detection by means of a sound spectrogram analysis. The paper presents a method of implementing neural networks on parallel, programmable computers, which can effectively address the computational requirements of such signal processing applications. The method is applicable to multilayer connectionist networks and two-dimensional, SIMD (single-instruction multiple data stream) processor arrays. A detailed description along with comparisons to previously proposed methods is provided for a mapping of a multilayer perceptron network with back-propagation learning algorithm. The mapping includes partitioning of inputs larger than the processor array. The performance of the method is evaluated using the Nettalk neural network and is compared to that of other methods. In particular, it is shown that the implementation of the method on the Systolic/Cellular machine of Hughes results in the processing rate equal to 100 MCPS.

I. INTRODUCTION

Neural networks, due to their adaptive, self-organizing, fault tolerant, and non-linear capabilities, are emerging as an attractive technology for a variety of signal processing applications, such as image and data compression (Iwata, Nagasaka et al. 1989; Luttrell 1989), communications (Paris, Orsak et al. 1989), radar target detection (Ahalt, Garber et al. 1989; Kwan and Lee 1989; Roth 1989), and sonar target detection (Gorman and Sejnowski 1988; Malkoff 1990). A specific example is our research on applying the back-propagation neural network (Rumelhart, Hinton et al. 1986) to sonar signal processing. In this application a sound spectrogram (Urick 1983) is used as the input to the neural network. The spectrogram is a three dimensional representation with one axis denoting time, one denoting frequency and the last axis denoting the underwater acoustic energy received by the hydrophone. The input pattern supplied to the neural network is constructed by extracting a small window of data from the spectrogram and performing application specific preprocessing on the data, see Figure 1. The output of the neural network represents the classification of the input pattern into one of the three categories : Target, non-target, or noise. The neural network model used in this application is a structured multilayer perceptron network with error back-propagation learning (Rumelhart, Hinton et al. 1986). The network is trained by repeated presentation of a training set with each training pattern being identified by a supervisor of being a target, non-target, or noise sample. The network *learns* by creating an internal representation for discriminating features in the training set through minimization of the mean squared error between its output and the correct response, supplied by the supervisor, to a given input training pattern. The sheer size of the spectrogram data and the training set size can drastically limit the practical use of this system on a conventional computer. The need for high throughput flexible architectures is apparent for implementations of neural networks for realistic applications.

Implementations of neural networks known in the literature span a full spectrum from software implementations on general purpose computers to strictly special-purpose hardware implementations. The software implementations are characterized by maximal flexibility but often inadequate processing speeds. The special-purpose hardware solutions, on the other hand, provide superior throughput at the cost of minimal flexibility. Implementations on programmable parallel machines, discussed in the paper, constitute a compromise between these two extremes. The architectures of these machines reflect better the structure of neural networks than those of sequential machines, and the available programmability provides the needed flexibility for efficient implementation of various neural network models on the same platform.

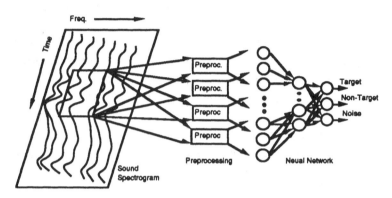

Figure 1 - Structure of a sonar target detection system utilizing neural networks

Neural network research is in its early stages of development and most of the progress is accomplished by means of experimental, rather than analytical studies. This fact has created a need for efficient and flexible implementations of these networks. Because of high cost of hardware design and prototyping, it may be impossible or impractical to develop new special-purpose devices each time existing models of neural networks are modified or entirely new models are introduced. Thus the programmability of hardware for neural networks may prove to be essential.

Among the existing and well understood parallel architectures, two-dimensional, mesh-connected, SIMD processor arrays stand out as a very natural class of machines for neural network implementations. The SIMD paradigm makes it possible to put together a very large number of processing elements without complicating the operating system or the programing of the machine. Moreover, the processing elements may be very simple, as the control and most of the memory management functions are performed by a shared, single controller. Parallel SIMD machines are also the best understood class of parallel machines. Many of them have been build and tested in laboratories (Little and

Grinberg 1988; Przytula and Nash 1987) and some have been developed as commercial products , e.g. Connection Machine (Hillis 1985) and DAP (Reddaway 1973). In this paper, we will focus our attention on implementation of neural networks on fine to medium grain SIMD arrays with mesh-connected processing units - i.e. each processor is connected to its four nearest neighbors. We assume also that the arrays have wraparound connections, but our mapping method can be easily modified to apply to the meshes without wraparounds.

In the context of this paper, the term mapping of an algorithm onto a parallel architecture is used to describe a method of finding a parallel version of the algorithm, here a neural network algorithm, and its realization on a parallel architecture. The mapping method has to take into account the following characteristics of the target architecture: topology of the processing array, granularity of the system, local memory size, communication vs. computation bandwidths, data formats, and computational capabilities of the processing unit (e.g. ALU or hardwired functional units).

The mapping method can be derived in two ways: automatic and heuristic. Automatic mapping is done by means of software tools, which take as input a given algorithm and an architecture description, and derive, often in an interactive way, the mapping of the algorithm onto the architecture. This approach was used to generate a mapping of neural networks onto a parallel architecture described in (Kung and Hwang 1989). Our mapping method as well as that in (Blelloch and Rosenberg 1987) and (Pomerleau, Gusciora et al. 1988) was produced heuristically, i.e. by trial and error, based on understanding of the algorithm and the architecture. This approach is usually more time consuming but often leads to a more efficient implementation.

In the derivation of the mapping, the target architecture may be specified only in general terms, so that the resulting mapping method is applicable to a class of machines rather than a single machine (Kung and Hwang 1989). A mapping method designed for a specific target machine, on the other hand, takes maximal advantage of its specific features but may not be very useful for other machines, even from the same class. Our approach has been to develop a general mapping for a class of mesh connected SIMD machines and to verify its performance on one example machine from that class, i.e. the Systolic/Cellular coprocessor recently completed at Hughes research laboratories (Przytula 1989; Przytula and Nash 1987).

The mapping method described in the paper is general also in the sense that it can be used for various classes of neural network models. In particular, the method is applicable to all the layered models in which operations can be represented as product of matrices and vectors, inner product of vectors, and arbitrary local operations. Where by local operations we mean operations which can be performed in unison in neurons or synapses without data transfers between them, such as calculation of the output transfer function. For more details regarding the characterization of neural network models from this point of view see (Kung and Hwang 1989). Our paper provides details for mapping of only the multilayer perceptron with back-propagation learning. This mapping, however, covers all the basic vector and local operations mentioned earlier. This is illustrated by observing that the recall phase of the multilayer perceptron involves matrix times vector and local operations, whereas the back-propagation learning is based on vector times matrix, inner product of vectors plus some local operations.

The paper consists of four sections. This introductory section is followed by a discussion of previous work by others in mapping neural networks onto programmable parallel architectures in section 2. A description of our mapping method is given in section 3. The details on the implementation of the Nettalk neural network (Sejnowski and Rosenberg 1987) on the Systolic/Cellular machine, using our mapping method, along with performance comparison to other implementations are presented in section 4. Finally, the conclusions are gathered in section 5.

II. OTHER MAPPINGS

Several different mapping methods have been proposed for mapping neural networks onto programmable parallel architectures. These methods can be roughly categorized into two groups, one for mappings onto specific parallel machines (fixed target architecture) and the other for mappings onto special-purpose architectures (i.e. customized target architecture). Several mappings of neural networks onto specific target machines have been described in the literature (Blelloch and Rosenberg 1987; Deprit 1989; Forrest, Roweth et al. 1987; Pomerleau, Gusciora et al. 1988; Witbrock and Zagha 1989). One such mapping is that of the multilayer perceptron with back-propagation learning onto the Warp machine of Carnegie Mellon University (Pomerleau, Gusciora et al. 1988). The Warp computer is a coarse grain, one dimensional, systolic, SIMD machine consisting of 10 powerful processors. The mapping requires processing of a complete network on each processor to utilize efficiently the processor's large local memory and computational power. In essence this mapping does not break up a single task (a neural network in this case) for parallel processing, rather it implements multiple independent networks concurrently. During the learning phase, the weights are updated in a batch mode (Rumelhart, Hinton et al. 1986), i.e. the updates are not done for each separate input pattern, but a sum of weight modifications corresponding to a batch of input patterns is applied.

Another interesting mapping onto a specific machine is the implementation of the Nettalk network on the Connection Machine CM1-16k (Blelloch and Rosenberg 1987). The Connection Machine CM1-16k is a fine grain SIMD machine consisting of over 16,000 bit serial processors with each processor having 64K bits of local memory. In the mapping of the multilayer perceptron network on this machine, a single network is distributed across the processors of the machine and executed in unison. Every neuron and every connection weight in the network is mapped onto a separate processor. The special *copy* instructions of the machine are used for efficient data movement and manipulation. The performance of this mapping on the CM1-16k, as well as the mapping on the Warp computer, are lower than our mapping onto the Systolic/Cellular machine, even though all of these machines have comparable throughputs.

There have also been a number of mappings onto special-purpose architectures proposed in the literature (Brown, Garber et al. 1988; Gaudiot, Malsburg et al. 1988; Kato, Yoshizawa et al. 1990; Kung and Hwang 1988; Kwan and Tsang 1990; Murry, Smith et al. 1989; Piazza, Marchesi et al. 1990). These mappings specify the system architecture, computation at each node, and the required communication between the nodes in order to implement a specific neural network model. The mapping described in (Kung and Hwang 1988), has been developed for the multilayer perceptron with back-propagation learning and a systolic, two-dimensional, mesh-connected processor array with wrap-around connections . In this method, each layer of the neural network is mapped onto a single row of the array. Input patterns are entered and processed in the array in a pipelined fashion, with the pipeline cycle time being equal to the amount of time necessary to process the largest layer of the network. Utilization of machine resources may be low for this method in case of a large discrepancy in the length of the layers of the network and also when the number of layers is much smaller than the number of rows in the processor array. We have evaluated performance of this mapping on the Systolic/Cellular machine of Hughes using the Nettalk network and it turned out to be much worse than that of our mapping under the same circumstances.

Many mapping methods, utilizing the same type of architecture, also suffer from low efficiency when there are discrepancies between the topology and the number of neurons in the neural network and the processor array. In particular, most mappings do not address the partitioning of the neural network over the processing array, which needs to be used when the number of neurons per layer exceeds the number of processors in a row of the processing array.

III. THE MAPPING METHOD

In comparison to the mappings described above, our mapping method is suitable for a large class of neural networks and a class of fine- to medium-grain, two-dimensional, mesh-connected, SIMD arrays. However, the method is described in detail for the multilayer perceptron network with back-propagation learning. The performance of the method is illustrated using the Nettalk neural network and the Systolic/Cellular coprocessor in order to allow for comparison with other implementations.

In our method, we assume that several input patterns, as many as there are rows of processors in the array, are available and are processed concurrently. This assumption is generally valid, especially during the training/learning phase of the network, when the complete training set is available before the start of processing. A single processor row of the array performs computation for a single input pattern as the processing propagates through all the layers of the network, see Figure 2. The method deals very effectively with partitioning of inputs larger than the processor row size and assures good performance due to the data organization and flow through the array.

In this section we describe the implementation of the data processing in the recall and learning phases of the multilayer perceptron (MLP) network with back-propagation learning (also called the back-prop model) on a 2-D mesh array architecture. The structure of the MLP network consists of several layers of neurons, where the first layer is called the input layer, the last is called the output layer, and all the layers between the input and output layers are called

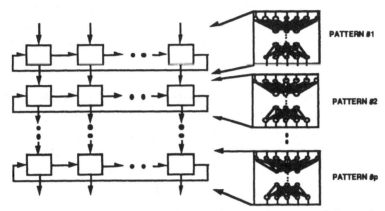

Figure 2 - Mapping of multiple neural networks onto a 2-D mesh-connected processor array

hidden layers. There are no synaptic connections between neurons on the same layer, but each neuron is generally connected to all other neurons in the adjacent layers.

The back-prop model operates in two distinct phases, one is the recall phase in which the training pattern is presented to the input layer of the network and a corresponding output is *recalled* at the output layer. The other is the learning phase in which the network adjusts its synaptic weights in order to minimize the error between the recalled pattern and the correct pattern supplied by a supervisor. The operations of each neuron in the the network during the recall phase are demonstrated in Figure 3, where *a* denotes the neuron activation value, *w* denotes the synaptic weight value, θ denotes the neuron threshold value, and *f* denotes the nonlinear neuron activation function. In short, the function of each neuron, during the recall phase, is to calculate the weighted sum of all its inputs and apply the nonlinear activation function to this sum in order to generate the neuron's output value.

Similarly, the processing involved in the learning phase is depicted in Figure 4, where *d* denotes the error term and *f'* is the first derivative of the neuron activation function. The error terms for neurons on the output layer ($\ell{=}L$) are calculated as

$$\delta_i(L) = f'(U_i(L))(t_i - a_i(L)) \tag{1}$$

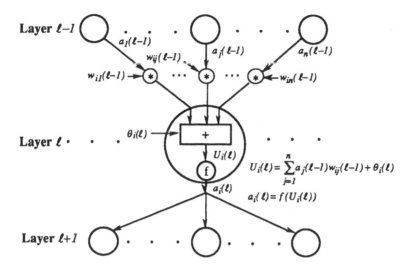

Figure 3 - Recall phase (forward pass) processing of neuron i on layer ℓ of the multilayer perceptron network model

where t_i denotes the desired output value of neuron i supplied by the supervisor. Whereas in the recall phase, the neuron activation values are propagated forward in the network, the error values are propagated backwards in the network during the learning phase.

In the remainder of this section we introduce a mapping method for efficient implementation of the operations in the recall and learning phases of the back-prop model on a 2-D mesh-connected SIMD architecture. For the sake of simplicity, let us assume for now that the neural network under consideration has the same number of neurons in each layer, and that this number is equal to the number of processors in a single row of the processor array. This assumption will be dropped later.

In our method, a different input pattern for the entire neural network is mapped onto each row of the processor array, as shown in Figure 2. Thus, each row processes data for one complete network starting from the first layer and then continuing sequentially for consecutive layers until the outputs are obtained. At the beginning of the recall phase, the processors contain, in their local memory, the inputs for the network, one input pattern per row and one item per processor, see Figure 5. The indices of the three different sets of

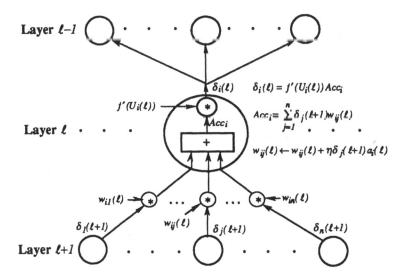

Figure 4 - Learning phase (error back-propagation) processing for neuron i on layer ℓ of the multilayer perceptron network

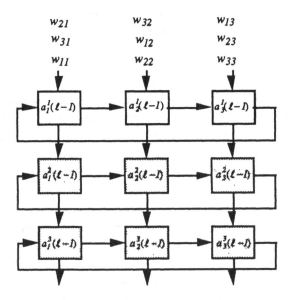

Figure 5 - Data organization in the processor array during the first cycle of the recall phase. Only processors in the first row are enabled.

inputs/activation values appear as superscripts, subscripts denote different neurons in a given layer, and the layer index is given in parentheses. The shaded squares represent the processors that are disabled (masked) in the given cycle.

The different input patterns are transformed into outputs of the first hidden layer of the networks by the processing involving vertical - North to South - flow of the synaptic weights w_{ij}, and horizontal flow - West to East- of the partial sums U_i . A single shift South is followed by a single shift West until the partial sums make a full circle. One processing cycle, for the jth processor, consist of the following operations: (1) receiving the weight value w_{ij} from the North neighbor and at the same time, sending the weight value w_{ij} received in the previous cycle to the South neighbor; (2) receiving the partial sum U_i from the West neighbor and at the same time, sending the old U_i to the East neighbor; (3) Multiplying the incoming w_{ij} by the a_j, which is stored locally, and adding it to the incoming U_i, see Figure 6.

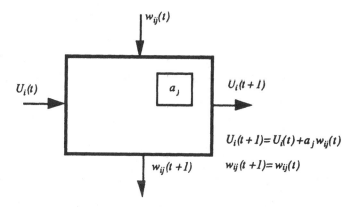

Figure 6 - Processing in each processor of the array during the recall phase.

The weight values are organized in a skewed fashion in the global memory. This organization is used in order to have the correct weight value w_{ij} and partial sum U_i meet in the same processor at the proper time. Figure 7 demonstrates the next two iterations following the initial cycle shown in Figure 5.

236

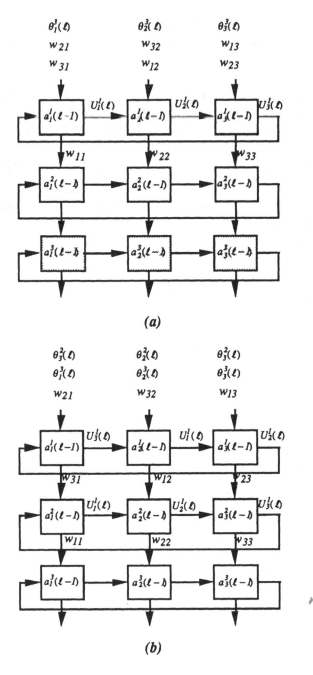

(a)

(b)

Figure 7 - Next two cycles of the data flow through the array after
initial cycle shown in Figure 5.

In this procedure, the vertical flow brings the same weights in contact with the different input patterns in the consecutive rows of the array. This is possible since the weight values are identical for all the networks in the array. Thus two levels of parallelism are achieved: the processing of complete networks is performed concurrently in a pipeline mode in the different rows of the processor array, and each network is processed in parallel by the different processors of each row. After all the weight values of a given layer have passed through the processor array and the threshold values q_i have been added to the partial sums U_i in each row, the computation of the activation function takes place in unison in all the processors. At the completion of this operation the activation values $a(\ell)$ for layer ℓ are available in each processor for computing the activation values for the next layer $\ell+1$. Concurrent with the propagation of weights through the array, the final U_i values in each row are propagated through the array and stored in the global memory for later use during the learning phase. This process is repeated until the activation values for the neurons on the output layer have been evaluated.

It is apparent from Figures 5 and 7 that the processing of multiple networks in the processor array is performed in a pipelined fashion. The organization of data in the memory is structured such that the pipeline flushing time is overlapped with loading of values for the next operation. This method is efficiently used to implement networks larger than the processor array by loading and unloading different segments of the network for processing.

During the learning phase, we also assign the computation of an entire network to each row of the processor array. This leads to a similar data flow and identical organization, in the array, of the activation and weight values. Thus the transition from the recall to the learning does not require data reorganization. The activation values a_i along with the accumulated sums Acc_i and the learning rate η are stored in each processor's local memory, see Figure 8.

The weights w_{ij} travel from North to South, as before, and the error values δ_i move from West to East. Two sets of weights are being transferred concurrently - the old and the new, where the new weights w_{ij} are computed on the fly according to the learning (or weight update) formula from Figure 9. The old weights w_{ij} are the weight values used during the recall phase. Treating the new and the old weights separately in this fashion insures consistency in learning between multiple networks executing in different rows of the processing array.

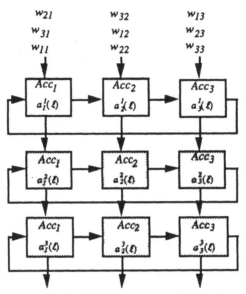

Figure 8 - Data organization in the processor array during the first cycle of the learning phase. Only processors in the first row are enabled.

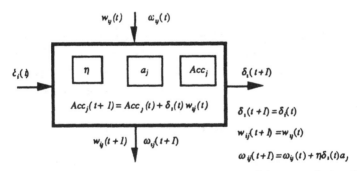

Figure 9 - Processing in each enabled processor of the array during the learning phase.

The processing produces modified synaptic weights w and the error values d for each consecutive layer of the network starting from the output down to the input layer. At each cycle the old weight value w_{ij}, received from the North port, is multiplied by the error term δ_i, received from the West port, and added to the partially accumulated sum Acc_j in each processor's local memory, see Figure 9. The error term δ_i is also multiplied by the neuron activation value a_j and the learning rate η, both stored in each processor's local memory, to calculate the weight modification value. The new weight value w_{ij}, received form the North port is updated by being added the weight modification value just produced in the processor. This procedure implements the generalized delta rule learning law (as shown in Figure 4). The next two cycles of the data flow following the initial cycle of Figure 8 are illustrated in Figure 10. At the completion of updating all the weight values between layer ℓ and $\ell+1$, the error term $\delta(\ell)$ for layer ℓ is calculated according to

$$\delta_i(\ell) = f'(U_i(\ell))Acc_i \qquad \text{for } \ell \neq L \qquad (2)$$

The accumulated sum Acc_i in (2) is calculated concurrently while updating the weights. The term $f'(U_i(\ell))$ is calculated by receiving the partial sum $U_i(\ell)$ values from the north port and implementing the f' function in unison all the processors. The partial sum values are those that where calculated in the recall phase and stored in the global memory for use here in the learning phase.

The neural networks used in practice have layers of different sizes and often contain larger number of neurons than there are processors in a row of a processor array (e.g. the sizes of each layer in our sonar target detection network are 365, 13, and 3 and the Hughes Systolic/cellular machine has 16 processors per row). Therefore, it is essential to address the issue of data partitioning as part of the mapping method. The partitioning in our method is implemented in such a way that the required local memory of the processors can be very small and independent of the neural network size. This is consistent with our assumption of fine to medium granularity of the processors as mentioned in the introduction.

If a given layer is smaller than the rows of the processor array, then some of the processors will contain phantom neurons with activation values always equal to zero. The processing in this case is not changed except that the weights between the real neurons and the phantom ones are set to zero. The layers that are larger than the rows of the processor array have to be processed in fragments of the size smaller or equal to the row size. The order in which these fragments are processed in the recall phase is shown in Figure 11.

240

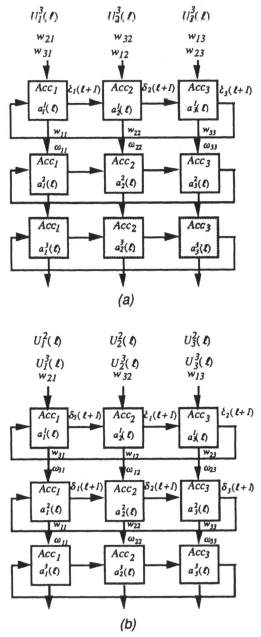

Figure 10 - Next two cycles of the data flow through the array after initial cycle shown in Figure 8. Shaded processors are disabled from performing calculations

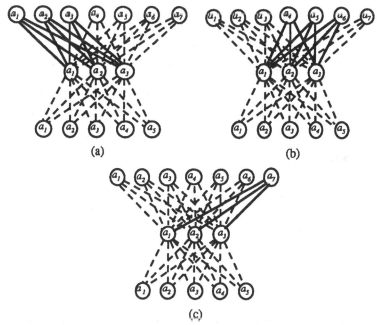

Figure 11 - Network partitioning for implementation on a "3 x 3" processor array

In this example the processor array size being used is "3 x 3" (similar to those in Figures 5 and 8) and there are 7 neurons on layer 1. The activation values from the first 3 neurons on layer 1 ($a_1^i(1)$ through $a_3^i(1)$, where i is the network index) are initially loaded into the processor array. The processing is performed, as described earlier, by propagating through the network the 9 weight values corresponding to the connections between neurons $a_1^i(1)$, $a_2^i(1)$ and $a_3^i(1)$ on layer 1 and the neurons $a_1^i(2)$, $a_2^i(2)$, and $a_3^i(2)$ on layer 2, during which the partial sum values U_i's will be accumulated. The activation values for the next 3 neurons on layer 1 ($a_4^i(1)$ through $a_6^i(1)$) are then loaded concurrently with the tail end of the processing phase of the last cycle, in a pipelined fashion. The next batch of 9 weight values for the connections between neurons $a_4^i(1)$, $a_5^i(1)$, and $a_6^i(1)$ on layer 1 and neurons $a_1^i(2)$, $a_2^i(2)$, and $a_3^i(2)$ on layer 2 are then propagated through the processor array as before and the computation of the partial sums continues. Finally, the last neuron on layer 1 $a_7^i(1)$ is loaded into the network and processed.

After all the neurons on layer 1 have been processed in this fashion and the threshold values been added to the partial sums, the neuron activation function is applied to calculate the activation values of the 3 neurons on layer 2 ($a_1^i(2)$, $a_2^i(2)$, and $a_3^i(2)$). If there were more than 3 neurons on layer 2, the activation values just calculated for neurons $a_1^i(2)$, $a_2^i(2)$, and $a_3^i(2)$ would be unloaded to the global memory and the above process would be repeated to evaluate the activation values for the next 3 neurons on this layer. This processing is repeated until the activation values for all the neurons on the output layer have been calculated. A similar partitioning procedure is used to perform the back-propagation of the error values through the network (learning phase). In this case segmentation starts from the output layer toward the input layer, as oppose to the reverse direction in the recall phase.

The pipelined processing employed in our mapping method allows for efficient loading and unloading of activation values from the processor array in this partitioning scheme. Since these operations almost completely overlap with the emptying and loading of the processing pipeline, only a small overhead results from processing the network in small partitions.

The execution procedure described in this section could be applied to many layered feed forward neural networks. Even though our mapping method uses a SIMD architecture, the activation functions used for the neurons can vary from layer to layer. Moreover, within each layer there is also a limited variability possible. For example if a polynomial approximation is being used to compute the neuron activation function, different polynomials can be implemented in each neuron at the same time by having different coefficients stored in each processor's local memory. If the network requires significantly different activation functions in the same layer, it can be accomplished by sequentially applying the desired functions and disabling/enabling appropriate neurons. Since the learning rate h is stored in the local memory of each processor, different neurons in the network could, if desired, use different learning rates with no effect on the performance of our mapping.

Neural networks with feedback architectures (such as Bidirectional Associative Memory (Kosko 1988) and Hopfield nets (Hopfield and Tank 1985)) can be implemented with the same type of architecture and style of mapping. The basic computation involved in the processing of these networks can be represented as matrix and vector calculations. The data organization and movement through the array, as in the mapping above, implement these operations very efficiently. The cellular operations in the algorithm, such as the activation function evaluation, could be changed to any other type of cellular operations without any effect on the inter-processor communication. This allows for great flexibility in the models that can be implemented with this type of mapping. In neural networks with feedback architectures, it might be more efficient to keep the synaptic weight values locally within each processor and transfer the activation values between the rows of the processing array. Our mapping method is most efficient for networks with dense interconnections between layers, although it could be used for networks having sparse interconnections just as well with a lower utilization factor. Since our mapping method takes advantage of the inherent regularity of most neural network models, it will not process efficiently neural networks with arbitrary and irregular interconnections.

IV. IMPLEMENTATION DETAILS

As described earlier, our mapping method is suitable for machines with 2-dimensional , mesh-connected, SIMD architectures. An example of such a machine is the Systolic/Cellular coprocessor designed and developed at the Hughes Research Laboratories (Przytula 1989; Przytula and Nash 1987). The architecture of this machine consists of a 16 by 16 array of processors controlled by a single controller in SIMD fashion, see Figure 12. The processors are connected to four of their nearest neighbors. Processors on the boundary columns are connected with each other through wrap-around connections. Each processor contains a small local memory (24 words) and seven 32-bit, fixed-point, functional units - two multipliers, two adders, a divider, and a comparator, which can all compute in parallel. A 2K words dualported data memory is used as a data queue. The dualported memory can be accessed in parallel by all the processors in the top and bottom rows of the array.

244

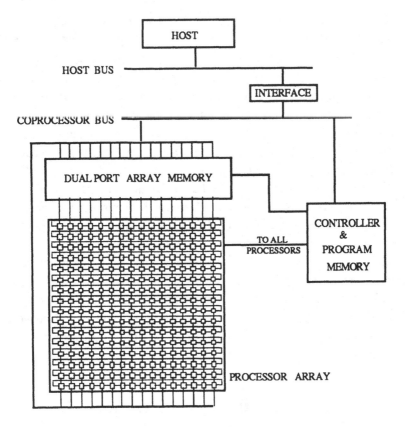

Figure 12 - System architecture of the Hughes Systolic/Cellular Coprocessor

We have tested our mapping method by implementing the well known Nettalk neural network (Sejnowski and Rosenberg 1987) on the Systolic/Cellular machine. The Nettalk network is a good example of a working neural network, which is large enough to require partitioning for our target machine and, as mentioned before, has been used by others to test their implementations of neural network mappings (Blelloch and Rosenberg 1987; Pomerleau, Gusciora et al. 1988).

The Nettalk network is a three layer feed-forward model which learns to pronounce written English text, through the use of the back-propagation learning algorithm. The network receives seven letters as inputs and the task of the network is to generate the correct phoneme corresponding to the middle letter. The input to the network is through 203 neurons, 7 groups of 29 possible input characters. The hidden layer contains 60 neurons and the output layer contains 29 neurons. Each layer is fully connected to its neighboring layers which yields a total of 13,826 connections in the network.

In our implementation, there were 16 input patterns executing concurrently - one per row of the processor array. Two neurons were mapped into a single processor to take advantage of the multiple functional units available in each processor. It is apparent from the mapping algorithm that there are approximately equal number of computation and communication operations. For example, the *double multiply* and the *double add* instructions were used in a single processor to multiply two activation values and their corresponding weights together and then add the results to the partial sum values in parallel. This system characteristic allows for efficient and simple overlapping of computation and communication operations. In our implementation of the Nettalk network on the Hughes Systolic/cellular processor, only 26.9% of the total processing time was attributed to the communication operations that could not be overlapped with computation.

Additional parallelism was exploited by overlapping computation operations with communication operations. For example, a North to South data transfer in all 16 processors in a row requires 3 cycles (when accessing global memory), and a multiplication operation takes 7 cycles. A multiplication operation was initiated in the multiplier and while this operation was being processed, the next two operands for the next instruction were fetched from the North processor concurrently.

The implementation of cellular operations, which are executed in unison across all the processors in the array, is not specified in our mapping. These operations have to be implemented by the most suitable method for a given target machine. A major portion of these operations are involves the computation of the neuron activation function. The activation function used in most back-prop implementations, including Nettalk, is the sigmoid function defined as

$$f(x) = \frac{1}{1 + e^{-x}} \tag{3}$$

In our implementation, this function was realized using the *exp(x)* function which in turn was calculated by means of a range reduction technique (Fike 1968). An alternative approach which may be more attractive for some machines is a look-up scheme. This scheme offers fast response (one memory read instruction), but has been rejected for implementation on the Hughes Systolic/cellular due to its idiosyncrasies.

Still another method for implementing the sigmoid is through numerical approximation. The amount of accuracy required for the approximation is dependent on the problem being addressed by the neural network. In our implementation the goal was to achieve an approximation with 16 bit accuracy. With this amount of accuracy, the approximation range to the sigmoid can be limited to between -10.5 and +10.5 and assume constant, 0.0 and 1.0 respectively, outside this range.

Several numerical approximations were explored. Since there is no local control in the processors of an SIMD machine, piece-wise approximation is not possible. Thus, a numerical method for implementing the sigmoid function was developed. Through a range reduction technique for calculating the $exp(x)$ function, we were able to arrive at an efficient implementation. The total number of operations required to calculate the $exp(x)$ is determined to be: 4 additions, 1 logical, 1 division, 1 shift, and 3 multiplications. To approximate the sigmoid function, using equation (3), an additional division and subtraction operations are needed.

In order to improve the execution speed of our implementation, we examined the use of other activation functions with similar properties to the sigmoid, but requiring fewer operations for evaluation. These functions need to be: nonlinear, monotonically increasing, differentiable, and bounded. One such function examined was

$$f(x) = \begin{cases} \dfrac{1}{1+x^2} & -\infty < x \le 0 \\ 1 & 0 < x < +\infty \end{cases} \qquad (4)$$

This function requires only 1 multiplication, 1 addition, 1 division, and 1 decision operation. Although it has a relatively more complex first derivative than the sigmoid, its implementation on the Systolic/cellular coprocessor requires only 24.3% of sigmoid cycles for the forward pass (recall phase) and 59.3% of cycles in the training phase (recall & learning), where both the function and its first derivative are used.

A measure of performance, which is becoming a standard for neural networks, is Million Connections Per Second (MCPS). The execution time of a complete recall and learning cycle, including all the data loading and unloading operations, was used to arrive at a 100 MCPS performance for our implementation of Nettalk on the Hughes systolic/cellular system. Considering only the recall processing we achieve a 240 MCPS performance with our implementation.

Table 1 combines the comparative results for our mapping method on the Systolic/Cellular Coprocessor, with mappings from (Kung and Hwang 1988) implemented on the same machine, and two other special purpose implementations on parallel machines - Warp (Pomerleau, Gusciora et al. 1988), and Connection Machine (Blelloch and Rosenberg 1987), and a workstation - Sun 3/160 with a floating point accelerator as reported in (Blelloch and Rosenberg 1987). The minimum local memory requirements for systems other than the Hughes systolic/cellular are estimates arrived at under the assumption that no loading/unloading operations from an external memory are allowed. The performance of the Connection machine implementation was also estimated using the performance reported in (Blelloch and Rosenberg 1987) multiplied by two to account for expected improvements due to code optimization. The implementation on the Warp used 32 bit floating point data with a floating point adder and multiplier and an integer ALU. Implementation on the connection machine used a two bit serial ALU per processor. The data type used in our implementation was 32 bit integer with integer arithmetic functional units.

	SYS/CELL HRL	SYS/CELL S.Y. KUNG	WARP	CM1 - 16K	SUN 3/160 + F.P.A.
PROCESSING RATE (MCPS)	100	18	17	7	0.034
MINIMUM LOCAL MEMORY (WORDS)	10	29	6182	5	14000

Table 1 - Comparison of different implementations of the Nettalk neural network based on throughput and local memory size requirements

The systolic/cellular prototype, used for our implementation of the Nettalk neural network, was implemented in older VLSI technology, with an 8MHz clock rate, and slow memory. The same machine implemented with current technology (1μ CMOS), with 25 MHz clock rate and fast memory devices, could achieve speeds of greater than 1 GCPS. A prototype of such a system is currently being developed at the Hughes Research Laboratories. This performance could be further improved with the use of easily implementable activation functions, such as the one in equation (4).

V. CONCLUSIONS

Neural network technology is being effectively applied to many signal processing applications. We have presented one such application - a sonar target detection system. It was shown that practical realization of such networks require a very high computational throughput. At the same time, the realizations have to be characterized by significant degree of programmability to accommodate the variability in the neural networks models. In this paper we have demonstrated a method for efficient mapping of neural network models onto programmable parallel machines. With this approach we can achieve performance comparable to expensive supercomputers, at much lower cost, and still maintain sufficient flexibility. The performance is also not much lower than that of strictly special purpose implementations and applies to large range of networks of different structure and size.

In particular, the mapping method is applicable to the feed-forward, supervised learning class of neural networks. A large number of neural network models fall into this this class. Most of them relay on the same basic neural operation but may very in interconnection structure and in some parameters of the learning algorithm. Also, some additional terms in the algorithms such as inertia (or momentum), although not discussed in detail here, can be easily implemented using the same mapping method. This applies to all extra terms that depend only on local information and therefore do not require additional data transfer steps.

The mapping method described in the paper is intended for implementation on fine or medium grain parallel machines, which are controlled in SIMD mode, and consist of processors interconnected into a rectangular mesh. Many of todays parallel commercial and experimental computers meet these basic assumptions, although they may very significantly in many aspects of implementation. The method provides especially good performance for architectures which provide sufficiently high I/O bandwidth to balance the computational power of the processors.

To illustrate the performance of the method, we have simulated a specific neural network on a selected parallel machine. A natural network of choice is Nettalk, which is an example of the multilayer perceptron with error back-propagation learning and has been used as a benchmark for other mapping methods and computers. Our target machine is the Hughes Systolic/Cellular Coprocessor. The implementation involved partitioning as the network is larger than the processor array. We have achieved performance of 100 MCPs (million connections per second) counting both the forward and backward propagation cycles.

The mapping method described in this paper provides capability of achieving high throughput by concurrently exploiting several levels of parallelism. The top level parallelism is due to the execution of multiple neural networks in different rows of the two dimensional processor array. The next level is due to processing of small parts of the network in multiple processing units in each row of the array. Two extra levels of parallelism could be added depending on the target computer. In particular, the data transfers between the processors could be overlapped with computations, and the computations could be distributed between multiple functional units (e.g. multiplier, adder etc) within each of the processors.

For example the Systolic/Cellular Coprocessors provides support for all these levels of parallelism with a very good balance between communication and computation. Approximately 54% of the processing time is spent on computation and 46% on communication. By implementing the mapping method with optimal use of multiple functional units and optimal overlap of computation and communication, the additional cost of communication can be limited to 26%.

Number of processor in the array is an important factor for the performance of the method. Although the method allows us to execute a network of arbitrary size on a given processor array, the processing time may be unnecessarily large if there is a very big mismatch between the sizes of the network and the array. For a small network implemented on a very large machine the losses come from presence of many phantom neurons which are disabled from processing. A very large network, on the other hand, may require partitioning into so many chunks on a small machine, that the loading and unloading costs may start dominating the computation.

We are continuing the work presented in the paper by extending the results to other classes of networks, such as recurrent networks. We believe that the modifications required in the method to accommodate the new models are going to be of a very minor degree.

REFERENCES

Ahalt, S. C., F. D. Garber, I. Jouny and A. K. Krishnamurthy. (1989). Performance Of Synthetic Neural Network Classification of Noisy Radar Signals. Advances in Neural Information Processing 1.

Blelloch, G. and C. R. Rosenberg. (1987). Network Learning on the Connection Machine. 10th Intern. Joint Conf. on Artificial Intelligence. 323-326.

Brown, J. R., M. M. Garber and S. F. Venable. (1988). Artificial Neural Network on a SIMD Architecture. Symposium on the Frontiers of Massively Parallel Computations. 43-47.

Deprit, E. (1989). "Implementing Recurrent Back-Propagation on the Connection Machine." Neural Networks. 2: 295-314.

Fike, C. T. (1968). Computer Evaluation of Mathematical Functions. Prentice-Hall.

Forrest, B. M., D. Roweth, N. Stroud, D. J. Wallace and G. V. Wilson. (1987). "Implementing Neural Network Models on Parallel Computers." The Comp. Jour. 30(5): 413-419.

Gaudiot, J.-L., C. v. d. Malsburg and S. Shams. (1988). A Data-Flow Implementation of a Neurocomputer for Pattern Recognition Applications. Areospace Applications of Artificial Intelligence.

Gorman, R. P. and T. J. Sejnowski. (1988). "Learned Classification of Sonar Targets Using a Massively Parallel Network." IEEE Trans. Accou., Speech, and Sig. Proc. **36**(7): 1135-1140.

Hillis, W. D. (1985). The Connection Machine. Cambridge MA, MIT Press.

Hopfield, J. J. and D. W. Tank. (1985). ""Neural" Computation of Decisions in Optimization Problems." Biol. Cybern. **52**: 141-152.

Iwata, A., Y. Nagasaka and N. Suzumura. (1989). A Digital Holter Monitoring System with Daul 3 Layers Neural Networks. International Joint Conference on Neural Networks. **2**: 69-74.

Kato, H., H. Yoshizawa, H. Iciki and K. Asakawa. (1990). A Parallel Neurocomputer Architecture Towards Billion Connection Updates Per Second. Inter. Joint Conf. on Neural Networks. **2**: 47-50.

Kosko, B. (1988). "Bidirectional Associative Memories." IEEE Trans. Syst., Man, Cybern. **18**: 49-60.

Kung, S. Y. and J. N. Hwang. (1988). Systolic Architectures for Artificial Neural Nets. IEEE Inter. Conf. on Neural Networks.

Kung, S. Y. and J. N. Hwang. (1989). "A Unified Systolic Architecture for Artificial Neural Networks." Journal of Parallel and Distributed Computing. **6**: 358-387.

Kwan, H. K. and C. K. Lee. (1989). Pulse Radar Detection Using a Multi-layer Neural Network. Inter. Joint Conf. on Neural Networks. **2**: 75-80.

Kwan, H. K. and P. C. Tsang. (1990). Systolic Implementation of Multi-Layer Feed-Forward Neural Network with Back-Propagation Learning Scheme. Inter. Joint Conf. on Neural Networks. **2**: 155-158.

Little, M. J. and J. Grinberg. (1988). The 3-D Computer: An Integrated Stack of WSI Wafers. Wafer Scale Integration. Boston, Kluwer.

Luttrell, S. P. (1989). "Image Compression Using a Multilayer Neural Network." Pat. Recog. **10**: 1-7.

Malkoff, D. B. (1990). A Neural Network for Real-Time Signal Processing. Advances in Neural Information Processing 2. 248-255.

Murry, A. F., A. V. W. Smith and Z. F. Butler. (1989). Bit-Serial Neural Networks. Neural Information Processing Systems. 573-587.

Paris, B.-P., G. Orsak, M. Varanasi and B. Aazhang. (1989). Neural Net Receivers in Multiple-Access Communications. Advances in Neural Information Processing 1. 272-280.

Piazza, F., M. Marchesi, G. Orlandi and A. Unicini. (1990). Coarse-Grained Processor Array Implementing the Multilayer Neural Network Model. Inter. Symp. on Cir. & Sys. **4**: 2963-2966.

Pomerleau, D. A., G. L. Gusciora, D. S. Touretzky and H. T. Kung. (1988). Neural Network Simulation at Warp Speed: How We Got 17 Million Connections Per Second. IEEE International Confer. on Neural Networks.

Przytula, K. W. (1989). Systolic/Cellular System.

Przytula, K. W. and J. G. Nash. (1987). A Special Purpose Coprocessor for Signal Processing. 21st Asilomar Conference on Signals, Systems and Computers.

Reddaway, S. F. (1973). DAP - a Distributed Array Processor. 1st Ann. Symp. on Computer Architecture. 61-65.

Roth, M. W. (1989). "Neural Networks for Extraction of Weak Targets in High Clutter Environments." IEEE Trans. on Sys., Man, and Cyber. **19**(5): 1210-12-17.

Rumelhart, D. E., G. E. Hinton and R. J. Williams. (1986). Learning Internal Representations by Error Propagation. Parallel Distributed Processing: Explorations in the Microstructure of Cognition. Cambridge, MIT Press.

Sejnowski, T. J. and C. R. Rosenberg. (1987). "Parallel Networks that Learn to Pronounce English Text." Comp. Sys. 1: 145-168.

Urick, R. J. (1983). <u>Principles of Underwater Sound</u>. McGraw-Hill.

Witbrock, M. and M. Zagha. (1989). <u>An Implementation of Back-Propagation Learning on GF11, a Large SIMD Parallel Computer</u>.

10

Implementation of Sparse Neural Networks on Fixed Size Arrays [1]

Manavendra Misra   V. K. Prasanna Kumar
SAL 344, Dept. of EE-Systems
University of Southern California
Los Angeles, CA 90089-0781
mamisra@pollux.usc.edu

Abstract

Recent research in Artificial Neural Networks (ANN's) has shown that ANN's will play an important role in solving many signal processing problems. To fully capture the potential that this new computational paradigm possesses, ANN models will have to be implemented in hardware. Initially, attempts were made to simulate ANN's on serial computers. These software simulations were too slow to be of any practical significance and it was realized that ANN's will have to be implemented on parallel machines that can exploit the parallelism inherent in ANN's. In this chapter, we investigate how sparse Neural Networks can be implemented on a fixed size mesh of processors. A number of currently available machines make available a computing environment based on this architecture and this architecture is also amenable to VLSI implementation. We show how one iteration of activation value updates for a sparse neural network with n neurons and e non-zero connections is simulated on a $p \times p$ array of processors in $O((n + e)/p)$ time. The efficiency of the algorithm is partly due to the fact that preprocessing is done on the connection matrix. This makes the algorithm efficient carrying out many iterations of the search phase computation with the same connection structure. Although not described here, learning algorithms like the Delta rule, which are based on the computation of a weighted sum, can also be run using a modified version of the algorithm.

[1] This research was supported in part by the National Science Foundation under grant IRI-8905929.

1   Introduction

The past few years have seen a resurgence in the field of Neural Networks. It has been realized that Neural Networks can provide a viable alternative to algorithmic processing for achieving brain like performance in a machine. The theory of Neural Networks has shown that Artificial Neural Networks (ANN's) can be used effectively in many Signal Processing applications [1, 5, 10, 34, 38].

Papers by John Hopfield in the early part of the eighties epitomized the resurgence in the field of Neural Networks. In a series of papers [11, 12], he borrowed techniques from Statistical Mechanics to show the power of a conceptually simple neural network model which has since been called the Hopfield Model in literature. The Hopfield Model consists of very simple processing elements interconnected to each other by weighted links. These neurons form a single computational layer with output signals fed back to the same layer. The network goes through a series of iterations till it converges on a steady output. Since the early eighties, there have been volumes of work to extend models that were developed in the fifties and sixties as well as research to develop new models to solve specific problems. The result of this research activity has been a series of models based on the neural paradigm which will find applications in various fields. Some of these models have theoretically been proved to be useful in Signal Processing applications. ANN's have been proposed for passive transient sonar signal classification, low frequency active acoustic sonar signal processing, adaptive beamforming and voice recognition amongst other applications [6, 7, 13, 15, 20, 22, 39].

A translation of ANN theory into application requires the implementation of these models. Initial attempts at implementation consisted of software simulation of ANN models on serial machines. These simulations further verified the usefulness of the models but were too time consuming to be of any major practical significance. It was realized by researchers working in this area, that the inherent parallelism of ANN's had to be exploited by implementations for them to make any significant impact. Parallel digital machines were therefore natural target architectures for these implementations.

The late eighties saw a series of articles describing research on the implementation of ANN's on parallel digital architectures. In [17, 18], S. Y. Kung and J. N. Hwang describe a scheme for designing special purpose systolic ring architectures to simulate neural nets. By recognizing that neural algorithms can be re-written as iterative matrix operations, the authors have been able to directly apply well known techniques for mapping iterative matrix algorithms onto systolic architectures. This method however, is efficient only for fully connected networks. Simulating sparsely connected networks requires the storing of zero weights for all the missing interconnections and unnecessary computations involving these weights. A considerable amount of space and time is thus wasted. Also, the existence of wrap-around connections is an undesirable feature of these architectures.

H. T. Kung et al [16] have simulated feedforward neural networks that employ the Backpropagation learning algorithm on the CMU Warp. The Warp is a programmable systolic machine with 10 powerful PE's and thus provides a coarse grain of parallelism. This coarse grain of parallelism makes it difficult to completely exploit the parallelism in all but the smallest of neural networks. Of the two algorithms described in the paper, the network partitioning scheme is inefficient for large networks, while the data partitioning scheme is effective during the learning phase but not in the search phase.

The Connection Machine is seen by many researchers as the perfect 'Neural Engine' because of its fine grain architecture. A simulation of multilayer ANN's running the Backpropagation learning algorithm on the Connection Machine CM-2 is presented in [44] by Zhang et al. The authors describe how to implement a *multiply-accumulate-rotate* iteration for a fully connected network, a process quite similar to the one described in [17], using the 2-D mesh connections of the CM-2.

Misra and Prasanna Kumar [25, 26] have presented algorithms to implement ANN's on the Reduced Mesh of Trees (RMOT) architecture. An RMOT of size n is an SIMD architecture with n PE's and n^2 memory modules arranged as an $n \times n$ array. The ith PE has access to memory modules in the ith row and i^{th} column through a set of row and column busses. The RMOT is shown to be an attractive architecture for implementing ANN's and algorithms are presented for implementing fully and sparsely connected single layer networks, as well as multilayer networks. An RMOT of size n is shown to perform as well as a $n \times n$ array of processors for this application.

Przytula et al [28, 29] describe algorithmic mapping schemes to map ANN models onto fine grained SIMD arrays. The implementations apply to connectionist networks of arbitrary topology in which search and learning operations can be expressed in terms of matrix and vector computations. The mapping methods developed are simple and general enough to be used on a number of commercial and existing machines. As a specific instantiation, the mapping methodology is shown for the Hughes SCAP machine.

A number of other researchers have also contributed to this rapidly developing research field that brings together Parallel Processing and Neural Networks. In [4], Scherson et al describe the implementation of neural network algorithms on the P^3 associative processor. Hammerstrom [8] has designed a digital VLSI chip for neural processing. Tomlinson et al [14] have used a different approach by using digital pulse trains to simulate biological pulse trains. Ramacher and Beichter [31] describe a modular systolic chip for emulating ANN's. Ranka et al [33, 32] have developed an ANN simulator for the Connection Machine and is working on a distributed implementation of Backpropagation on a LAN. Shams and Przytula [37] present a method for mapping multilayer ANN's onto 2-D SIMD arrays. Tomboulian [40] uses a method developed to route arbitrary directed graphs on SIMD architectures to simulate ANN's. Wah and Chu [41] describe a mapping methodology for mapping ANN's onto multicomputers. Watanabe et al [42] and Wilson [43] present ways to implement ANN's on specific array processors.

As is evident from the references given above, a number of researchers have tried to implement ANN models on arrays of processors. Amongst these, the linear array is the simplest. A number of real machines based on this architecture exist and thus, it is important to develop efficient application algorithms for this architecture. However, if the application in question requires a large amount of data to be transferred between processors, the limited bandwidth of communication in the linear array becomes a bottleneck and it becomes impossible to develop efficient algorithms. For such applications, better performance can be achieved on an architecture in which the processors are connected in the form of a two dimensional mesh. The 2-D mesh provides higher bandwidth than the linear array without increasing the complexity of interconnection significantly. Thus, it is still possible to implement a mesh connected computer in VLSI.

The implementation of sparse neural networks on parallel machines is one such application that requires a large amount of data transfer. In this chapter, we show how sparse neural networks can be implemented efficiently on a mesh connected computer of fixed size. We use a fixed size processor array of size $p \times p$ to implement the search phase computations of a sparse ANN. A network with n neurons and e non-zero connection weights is implemented on the array in $O((n + e)/p)$ time. Each PE has local storage of size $O((n + e)/p^2)$. Preprocessing on the connection weight matrix is necessary to generate an efficient data routing strategy. Ideas from Interconnection Networks are used to show that this routing can be done in a conflict-free manner.

The rest of this chapter is organized as follows: Section 2 describes the neural models addressed by this chapter. Section 3 describes the issues involved in implementing sparse ANN models on processor arrays and then presents a description of the target architecture, a 2-D array with p^2 processors. Section 4 describes the algorithms to implement sparse ANN's on the target architecture. Section 5 concludes the chapter.

2 Neural Network Models

This section describes the biological model of a neuron and then shows how its salient computational properties are abstracted to form the computational model. A general model that encapsulates the important properties of most ANN models is presented. Finally, in the last subsection, a brief insight into the learning mechanisms incorporated in ANN's is presented, with special emphasis on the Backpropagation Algorithm.

2.1 Biological Inspiration

The animal brain is a large conglomeration of very richly interconnected simple processing elements called neurons. Typically, the nervous system has about 10^{11} neurons, each having an average of 10^3-10^4 inputs and outputs giving rise

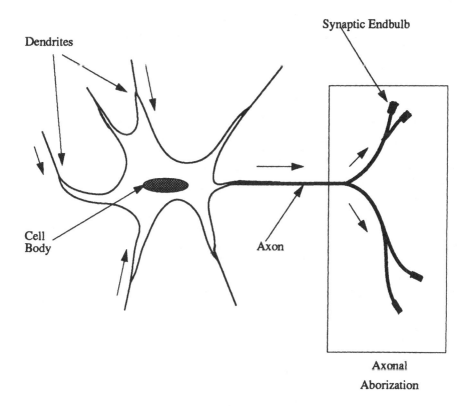

Figure 1: The Biological Model of a Neuron.

to 10^{15} interconnections. It is theorized that the immense computing power of the brain is a result of the parallel and distributed computing carried out by these neurons.

There is a great variety in the structure of neurons found in the nervous system and if looked at microscopically, neurons can be very complex. The biological model of the neuron however, captures all the salient features of a real neuron in a simple model (Figure 1). Dendrites form the input channels of a neuron while the axon forms the output channel. Axons of other neurons impinge upon the dendrites of a neuron through junctions called synapses. Synapses can be either excitatory or inhibitory and have weights associated with them. Signals are passed electrically through axons and then are transmitted chemically across synapses. A weighted sum of all the signals being received by a neuron's dendritic structure is formed at the cell body and this determines the membrane potential of that cell. The output of a neuron is a train of pulses sent out on the axon. The magnitude of these pulses remains

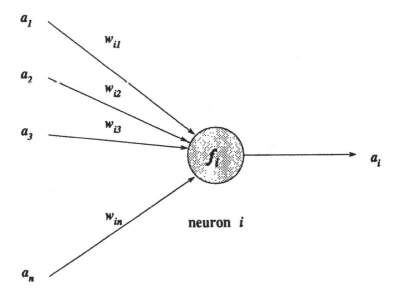

Figure 2: The McCulloch-Pitts Neuron

constant and information is conveyed in the firing rate of the neuron. A continuous monotonically non-decreasing function relates the membrane potential to the firing rate of the neuron.

2.2 Model of the Neuron

The computational model of the neuron used in ANN's is an abstraction of the characteristic properties of the biological neuron. The earliest neural model was developed in the 1940's by McCulloch and Pitts (Figure 2).

The McCulloch-Pitts Neuron [23] is a simple two state device. It forms a weighted sum of its inputs and yields a binary output depending on whether the weighted sum is greater than or less than a threshold θ.

$$a_i = \begin{cases} 1 & \text{if} \quad \displaystyle\sum_{j=1}^{n} w_{ij} a_j > \theta_i \\ 0 & \text{if} \quad \displaystyle\sum_{j=1}^{n} w_{ij} a_j < \theta_i \end{cases} \tag{1}$$

where a_i is the activation of the ith neuron, w_{ij} is the weight of the connection from neuron j to neuron i and θ_i is the threshold of the ith neuron. To mimic the biological model more closely, this transfer function could be replaced by a continuous, monotonically non-decreasing function which better matches biological data. One such function that is often used is the Sigmoid function:

$$S(x) = \frac{1}{1 + e^{-cx}} \tag{2}$$

where c is a constant. Neurons with continuous transfer functions are called Graded Response Neurons [12].

2.3   The General Model

A number of ANN models have been proposed in literature [21]. These models can be differentiated on the basis of:

- Whether the network is a single or a multilayer network.

- Whether it is a feedforward network or it has feedback.

- Whether the network incorporates learning [1] or not.

The computations involved in most ANN models however, conform to a common form. The neural networks addressed here adhere to the following general model. A neural network consists of interconnected simple neurons. The input signals received by a neuron are multiplied by appropriate weights and summed to yield the overall input to the neuron. The output of the neuron is produced by applying a function f, called the activation function, to the weighted sum.

The update step can be formally described as

$$a_i^{k+1} = f_i(\sum_{j=1}^{n} w_{ij} a_j^k) \tag{3}$$

The neurons in the network could form a single layer with feedback connections or could form the input, output and hidden layers of a multilayer network. In a single layer network, a neuron computes its new activation value after receiving inputs from other neurons, and communicates the updated activation

---
[1]Learning is defined to be the updating of synaptic weights.

value to neurons its output connects to. In a multilayer network, the activation values are communicated to the next layer. A forward pass of data from the input layer to the output layer, which does not involve changes of weights, is referred to as a *recall operation* or the *search phase*. Learning can either be executed in the forward pass by carrying out additional computations in the neurons or may require a separate pass of data in the opposite direction (as in the Backpropagation model).

2.4 Learning

Learning is defined as the modification of synaptic weights that encodes patterns into the ANN. Learning can either be supervised or unsupervised. In unsupervised learning (eg. Hebbian Learning [9]), the weight of a link is updated based on local information available to the neurons connected by the link. Supervised learning [35], on the other hand, requires the presence of an external "teacher". The teacher modifies the weights based on the error between a desired response and the actual response to an input.

One of the most popular learning schemes for multilayer neural networks is the Backpropagation Algorithm [36]. Backpropagation is a supervised learning mechanism which minimizes the mean squared error between the desired and actual output values. One of the reasons that it is often used to solve real life problems is that it is computationally cost effective. There are two phases to the Backpropagation Algorithm. In the forward pass, the training pattern is input to the network and activations of the neurons are updated till the output emerges at the output layer. This output is compared with the desired output for that pattern and the error signals are propagated back through the network and the weights are updated. The computational complexity of the backward phase is the same as that of the forward phase.

At the start of a training run, a training pattern is input to the input layer of the multilayer ANN. Let the actual output of the jth neuron of the output layer be a_j. Let the desired (or target) output at that neuron be t_j. Then, $(t_j - a_j)$ defines the error ϵ_j at that neuron. In general, the change in weight w_{ij} is given by:

$$\Delta w_{ij} = \eta \delta_i a_j \tag{4}$$

where w_{ij} is the weight of the connection from neuron j to neuron i, η is the learning rate and δ_i is the error signal. The error signal is defined as follows. If neuron j is an output unit, then:

$$\delta_j = \epsilon_j f_j^{'}(x_j) \tag{5}$$

x_j is the weighted sum of inputs to neuron j and $f_j^{'}$ is the derivative of the activation function. The error signals for the hidden units are computed recursively:

$$\delta_j = f_j^{'}(x_j) \sum_k \delta_k w_{kj} \tag{6}$$

It should be noted that the major computation in the Backpropagation algorithm requires the computing of a weighted sum (Equation 6) which makes the structure of the computation very similar to that of the search phase.

3 Neural Network Implementation on Processor Arrays

Implementing sparse neural networks on parallel architectures involves a large amount of information to be transferred between the processors. Although the linear array of processors is an attractive architecture because of its simplicity and availability in terms of real machines, it is not efficient for simulating sparse neural networks because of the restricted bandwidth of communication it possesses. The 2-D array of processors alleviates this problem as it has a higher bandwidth. In this section, we first briefly describe S. Y. Kung and J. N. Hwang's implementation of a fully connected neural network on a linear array with a wrap around connection, identify its shortcomings and then describe the target architecture that will be used in this chapter.

3.1 ANN Implementation on Linear Arrays

The processing required to carry out one update of activation values for the general ANN models described in Section 2 can be seen to be a matrix vector multiplication. Recognizing this, S. Y. Kung and J. N. Hwang have applied techniques for mapping iterative matrix operations onto systolic arrays to develop algorithms to simulate ANN's on linear arrays [17, 18]. The architecture used for this implementation, along with the initial distribution of activation values and interconnection weights is shown in Figure 3.

The array consists of n PE's (where n is the number of neurons in the ANN) which are connected to form a linear array with a wrap around connection. Each PE is initially assigned one activation value and has access to a local memory that stores all the weights associated with the corresponding neuron, rotated in the manner shown in Figure 3. The working of the array in the search phase is as follows. At any time step k, the activation values a_j^k are circulated clockwise around the ring. When a value a_j arrives at a PE i, it is multiplied by w_{ij} and added to a partial sum. After $O(n)$ time, all the products have been added. The activation function f is applied to this resulting sum to get a_i^{k+1}. This process is executed repeatedly till the network has converged to a stable state.

The above method is efficient for a small, fully-connected network (notice that $\Omega(n)$ time is required for implementing a network with n neurons). If, however, the network being implemented is large and sparsely connected, the above method is very inefficient. The method stores all values of weights, whether or not they happen to be zero. For a large, sparse network, a large proportion of the weights could be zero. The above method would still store all n^2 weights and carry out all computations. In such cases, only the non-zero weights need

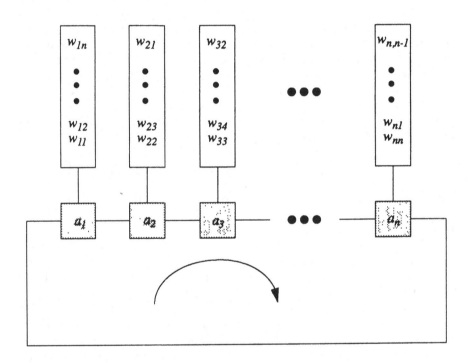

Figure 3: Linear Array implementation of ANN's proposed by Kung and Hwang.

to be stored and computed with. This results in a saving of memory and a more efficient algorithm. Also, the above algorithm requires a problem dependent array size of n PE's. Real machines have a fixed number of processors and it is important to develop algorithms for cases when the number of processors is smaller than the problem size (which would typically be the case).

Although it is possible to store only the non-zero weights on a linear array, this requires a considerable amount of data transfer between PE's. It turns out that the linear array is not a very suitable architecture for this application because its limited communication bandwidth makes this data transfer expensive. Here, we present an efficient algorithm to implement sparse neural networks on a fixed size mesh of processors where it is possible to store just the non-zero weights and carry out the data transfers efficiently. A description of the target architecture is presented next.

3.2 The Target Architecture

The organization of the array of processors used in this chapter to implement ANN models is shown in Figure 4. The Architecture consists of p^2 processing elements (PE's) connected to form a $p \times p$ two dimensional array. The architecture operates in an SIMD mode and therefore requires a host computer to control the operation. Communication between PE's is bidirectional and each PE can communicate with one of its four nearest neighbors in one time step. The bidirectional links are used to carry out the data routing in the array. Each PE has a local memory of size $O((n + e)/p^2)$ (where n is the number of neurons in the network and e is the number of non-zero interconnection weights), a multiplier, an adder, a lookup table for computing the activation function, and the ability to communicate with its neighbors. For purposes of computing the time complexity of the algorithms in this chapter, we assume that one time step is either a computation or a communication step.

The appeal of the architecture described above lies in its simplicity and its availability in the form of existing parallel machines. It is also possible to build a custom designed parallel machine based on this architecture by using off the shelf components. In the next section, we present the algorithm to implement the search phase of sparse ANN models on this target architecture.

4 Fixed Size Arrays for Sparse ANN Implementation

This section describes an efficient means of implementing sparsely connected neural networks on a fixed size 2-D array of processors. Before presenting the algorithm for simulating the search phase of sparse ANN's, we show how the non-zero connection weights and activation values are mapped onto the PE's of the mesh.

 **Processing Element**

Figure 4: Two Dimensional Array of Processing Elements

4.1 Mapping of Weights and Activation Values

We assume that the ANN to be simulated has n neurons and e non-zero connections. As was described in Section 3, the target architecture is a $p \times p$ mesh of PE's. Thus, there are $(n + e)$ data elements to be stored between the p^2 PE's and each PE stores $O((n + e)/p^2)$ values. To better visualize the storage of data on the mesh and its processing, we can imagine the data registers forming a set of $(n + e)/p^2$ 'planes' of registers (Figure 5). Each of these planes of registers holds p^2 data values. The mapping of the data onto the memory registers is done such that data routing can be done on a plane-by-plane basis, routing within each plane being equivalent to a permutation of the elements within that plane. This visualization helps in understanding the working of the algorithm and in the analysis of its complexity.

The scheme used for mapping data onto the array is a direct result of the data routing requirements arising from the computation structure. Recall that each update of activation values comprises of the following computation:

$$a_i^{k+1} = f_i(\sum_{j=1}^{n} w_{ij}a_j^k) \qquad \text{for } 1 \le i \le n$$

This computation requires that a PE that stores the activation value a_j^k at the end of iteration k, has to route this value to all PE's that contain non-zero weights from column j of the connection weight matrix $\mathbf{W}$. After the activation values are multiplied with the weights, the products that correspond to rows of $\mathbf{W}$ are summed. These weighted sums are then routed back to the PE that stores the appropriate activation value. Upon receiving the weighted sum, the PE applies the activation function on this value to get the new activation value, a_j^{k+1}.

Mapping of Activation Values: As mentioned above, the routing scheme presented in this paper perceives the routing of data within a plane as a permutation of data within that plane. Using concepts from the field of Interconnection Networks, we show that it is possible to simultaneously route all the data elements within a plane in a conflict free manner. The first step of the computation requires that activation values be routed to PE's that contain non-zero weights from the appropriate columns of $\mathbf{W}$. Since all the elements in a plane can be routed simultaneously, the activation values to be routed during this first step should be spread out over the plane. Thus, activation values are mapped row by row onto the first plane, starting from the top left corner register of the plane (Figure 5). If there are more than p^2 activation values, more planes may be needed. The mapping within each subsequent plane is similar to the first plane.

Initial mapping of Connection Weights: The above computation dictates that all PE's that store non-zero weights from column j of $\mathbf{W}$ must receive activation value a_j at the end of the first routing stage. A natural deduction from this is that weights belonging to the same column should initially be stored as close to each other as possible. This will make the distribution of a_j amongst

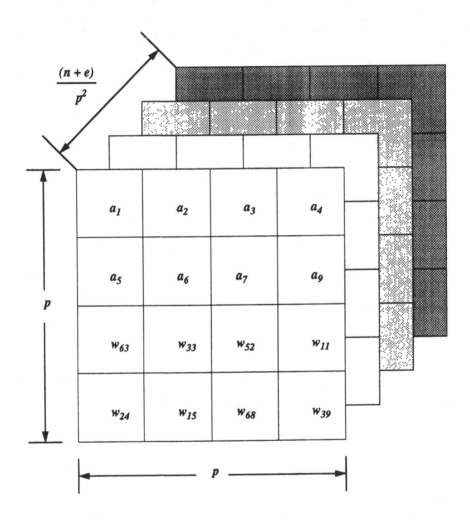

Figure 5: Stored data can be looked upon as being stored in planes of registers

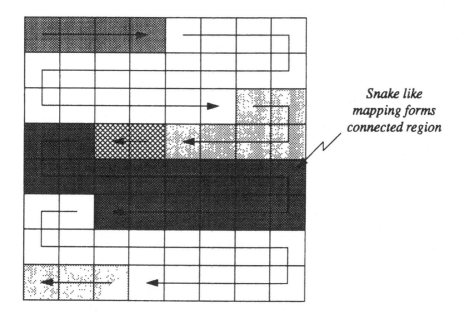

*Snake like
mapping forms
connected region*

Figure 6: Data sets mapped onto a 2-D array form connected regions if the mapping is done in a 'snake-like' manner. Each shaded region represents a data set which forms a connected region with respect to the 2 D mesh connections.

them easier. Thus, we first map as many weights from one column as possible onto one PE. If more memory is needed than is available in one PE, subsequent PE's are chosen such that the PE's form a 'connected region' with respect to the physical connections of the mesh. On a 2-D mesh, data sets form connected regions if they are mapped in a 'snake-like' manner onto the array (see Figure 6). Thus if it is necessary to store weights from one column over many PE's, it is done in a snake-like manner, filling each PE completely before going on to the next. We shall define such a mapping to be a 'snake-like column major order' mapping. Note that weights from one column can transcend many planes.

Later during the algorithm, it will be necessary to sum all the product terms that correspond to one *row* of **W** and then, it will be required that these product terms form a connected region. This leads to a 'snake-like row major order' mapping of the product terms. Plane by plane data routing will be used to transform the initial snake-like column major order mapping to snake-like row major order mapping.

4.2 Data Routing

Before a description of the complete algorithm, let us consider the general problem of routing data within one plane of registers. During each iteration of the search phase, three kinds of data routing problems arise:

1. The broadcast of a_j to all elements of the jth column of the connection weight matrix.

2. Transformation from a snake-like column major order distribution to a snake-like row major order distribution.

3. Transportation of the sums $(\sum_{j=1}^{n} w_{ij} a_j^k)$ to the PE's that store the activation values.

We develop an efficient method that uses preprocessing done on the structure of the weight matrix to simultaneously route all the data within a plane in $O(p)$ time [24]. We describe the technique to be used to route data before we look at how each individual routing problem is solved.

The problem of data-transport among registers in a plane is essentially that of realizing a permutation of the elements contained in the registers. More formally, if register R_{ij} has to send data D_{ij} to $R_{i^* j^*}$, then the permutation to be realized is $\pi : (i, j) \rightarrow (i^*, j^*)$. An approach towards realizing such a permutation is to apply the following two steps. In the first step, the elements are moved within their columns till they are in their respective destination rows. In the second step, the elements are moved within their rows till they are in their destination registers. This method, however, could result in many elements accumulating in one register at the end of the first step (eg. if all the elements of a column have the same destination row, they will all end up in the same register). To avoid this kind of congestion, the elements are first permuted within their rows in such a manner that when the permutations along the columns are carried out, no two elements end up in the same register [30].

The 'three-phase' routing method can therefore be described as follows:

Phase I :
Permute the elements within their rows so as to avoid congestion in Phase II.

Phase II :
Permute the elements within columns so as to get them to their destination rows.

Phase III :
Permute the elements within their destination rows so as to get them to their final positions.

Three-phase routing of data is pictorially represented in Figure 7, which can be identified as the Clos Interconnection Network [2], [3].

The rectangular boxes in the figure represent a permutation of a particular row or column in the plane. For the above routing scheme to be able to realize any permutation of elements, each of the row and column sub-permutations should be able to realize any desired rearrangement. This is indeed possible in $2p$ steps (for a row permutation: p steps are sufficient to move data that have destinations to their right and p steps are sufficient for left movements). The overall routing process therefore takes $6p$ steps.

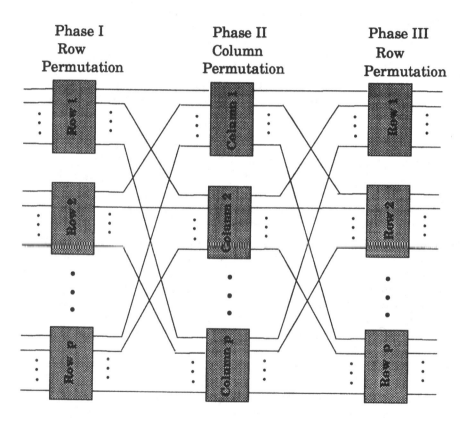

Figure 7: Permuting the contents of the PE's

Each phase of routing requires one routing register per data element which contains the distance the element has to be moved in that row/column and in which direction. We shall assume that positive numbers represent movement to the right (downward movement for a column) while negative numbers represent movement to the left (upward movement for a column). Each piece of data therefore has three routing registers associated with it: RR_I, RR_{II} and RR_{III}. The computation of the routing tags of data in one plane can be done in $O(p^2[\log p]^2)$ time on a serial computer [19] and in $O(p)$ time on a $p \times p$ mesh connected computer [27].

Using the above concepts, we develop a procedure called PHASE which instructs each PE to move data in parallel to one of its nearest neighbors depending on the phase being run. Another procedure, ROUTE, calls PHASE three times, once for each phase of the routing. Details of these procedure can be found in [24]. Using these procedures, the broadcast, re-distribution and update problems are solved as follows.

A. BROADCAST

At compile-time, given the sparse connection matrix ($\mathbf{W}$), the mapping of the non-zero elements onto the memory array in snake-like column major order is computed. This gives the connected regions in the array. An element of $\vec{a}$ has to be broadcast to each of these connected regions. To do this, a *leader* PE is identified in each connected region and the element of $\vec{a}$ is transported to it using the procedure ROUTE on the planes that contain activation values. If the leader PE of two regions coincide, a pseudo leader is chosen for one of the columns and the corresponding activation value is routed to that PE. This part of BROADCAST is completed in $6p$ time steps. The leader PE is either the leftmost or the rightmost PE of the top row of the connected region. The leader PE has two limit registers associated with it, one indicating how far the connected region extends along its row and the other showing how far down along its column the region extends. The data in the leader PE can be copied into all the PE's in the connected region in three stages. In the first stage, the data is moved to PE's in the same column within the connected region. In the second stage, data is broadcast to connected PE's in the same row and finally, a third stage broadcasts to connected PE's in the same column again. Each stage can take up to $2p$ time and so the total time required for copying data within a connected region is $\leq 6p$. The procedure BROADCAST, therefore, consists of calling the procedure ROUTE on the planes containing elements of $\vec{a}$ and then carrying out the data movement described above. The time required for one such broadcast operation is $\leq 12p$.

After the PE's receive the appropriate elements of $\vec{a}$, the PE's compute the products $w_{ij} * a_j$.

B. RE-DISTRIBUTE

The problem can be stated as follows:

Given a distribution of elements in snake-like column major order, re-distribute them in snake-like row major order.

At compile-time, the sparse connection weight matrix ($\mathbf{W}$) is available. The mapping of the non-zero entries in snake like column major order as well as in snake like row major order can therefore be computed. The routing problem is then reduced to realizing a plane-by-plane permutation of the contents of the PE's that converts the distribution from one to the other. This permutation is realized using the three phase routing technique described above. If, during the preprocessing, it is determined that two or more data elements have the same destination register as part of the routing within a plane, the ordering of data in a PE can be changed to avoid this.

The procedure RE-DISTRIBUTE is quite similar to ROUTE. Due to the similarities of the procedures, details are omitted. The time required for RE-DISTRIBUTE is $6p$ per plane. Since there are $O((n+e)/p^2)$ planes, this leads to a total time of $O((n+e)/p)$.

C. UPDATE

The solution of this problem is similar to the solution of the broadcast problem. During preprocessing, when the snake-like row major ordering of the non-zero entries is computed, the leader PE's of the connected regions so formed are also determined. The sums of the connected regions come to these leader PE's and so it is known beforehand which PE's have to send data to the PE's storing the components of $\bar{a}$. The routing registers of these PE's are set accordingly. The procedure UPDATE first forms the weighted sums and gets them to the leader PE's. These weighted sums are then routed to the appropriate PE's using ROUTE and the components of $\bar{a}$ are updated by the application of the function f. It is easy to verify that the time required for UPDATE is $\leq 12p$.

4.3 Complete Algorithm for the Search Phase

The complete algorithm to update activation values of the neurons of a sparsely connected neural network is presented in this section. The procedures described in Section 4.2 are used in the algorithm. In the pre-processing stage shown in Figure 8, the processor array is set up to perform the iterations. The iterations are performed as shown in Figure 9. The analysis of the time performance of the procedures in Section 4.2 shows that the complete algorithm runs in $O((n+e)/p)$ time.

The algorithm presented in this chapter is general enough to handle any computation that requires the repeated multiplication of a sparse matrix with a vector. Since a number of learning mechanisms like the Delta Rule and Backpropagation can be re-written as matrix-vector operations, the algorithm can be modified to implement them on fixed size arrays too.

1. Compute the mapping of the non-zero elements of **W** onto the processor array in snake like column major order and store it.

2. Identify the 'leader' PE's in the connected regions formed in Step 1. Set the limit registers in these PE's to define the boundaries of the connected regions.

3. Compute the mapping of the non-zero elements of **W** onto the processor array in snake like row major order and store it.

4. Identify the leader PE's in the connected regions formed in Step 3. Set the limit registers in these PE's to define the boundaries of the connected regions. Set the routing registers of the PE's that correspond to the update step so as to get these data to the PE's storing the activation values.

5. Store initial activation values a^0 in the manner defined in Section 4.1. Compute the routing tags to route these elements to the leader PE's identified in Step 2 and store them in the routing registers corresponding to the broadcast step.

6. Map the non-zero elements of **W** onto the array in snake like column major order according to the mapping computed in Step 1. Compute the routing tags for these PE's so as to achieve a permutation from the mapping in Step 1 to the mapping in Step 3.

Figure 8: Preprocessing Steps

repeat

1. Broadcast the activation values using the procedure BROADCAST. Routing registers determine which PE's the values go to. The products of the activation values and the weights are computed.

2. Use RE-DISTRIBUTE on each plane of registers to transform the distribution to snake-like row major order.

3. Use UPDATE to add the products and route them to the PE's storing the activation values. These PE's then update the activation values by applying the activation function on the weighted sum.

until (the network converges).

Figure 9: Iterations to update Activation Values

5 Conclusion

The two dimensional array of processors has a simple interconnectivity pattern that makes it amenable to VLSI implementation. A number of real machines based on this architecture are currently available. Also, it is possible to construct a custom designed parallel machine based on this architecture by using off the shelf components. Thus, it is important to develop efficient algorithms to solve problems on a two dimensional mesh of processors of fixed size. The two dimensional connectivity provides a higher bandwidth of communication as compared to the limited bandwidth available on the linear array. This makes the two dimensional array a better target architecture for applications that require a large volume of data to be transferred between processors.

One such application is the implementation of sparse neural networks. This chapter has presented an efficient algorithm for carrying out the search phase computations of a sparse neural network on a fixed size mesh of processors. One iteration of activation value updates for a network with n neurons and e non-zero connection weights takes $O((n + e)/p)$ time on a $p \times p$ array of processors. Each PE of the array contains a multiplier, an adder and $O((n + e)/p^2)$ local storage. Preprocessing is carried out on the connection weight matrix which results in efficient data routing within the array. Concepts from the field of Interconnection Networks are used to prove that routing can be done in a conflict free manner.

The algorithm presented in this chapter is general enough to be applied to any sparse neural model where the search or learning phase computation can be described as a product of a matrix with a vector. Since a number of neural models fall into that class, our method will be effective for all of them.

References

[1] R. Altes. Unconstrained minimum mean-square error parameter estimation with Hopfield networks. In *International Conference on Neural Networks*, volume II, pages 541–548, 1988.

[2] V. E. Benes. On Rearrangeable Three Stage Connecting Networks. *B.S.T.J.*, 41:117–125, Sept. 1962.

[3] C. Clos. A Study of Non-Blocking Switching Networks. *B.S.T.J.*, 32:406–424, 1953.

[4] K. I. Diamantara, D. L. Heine, and I. D. Scherson. Implementation of neural network algorithms on the P^3 parallel associative processor. In *International Conference on Parallel Processing*, volume I, pages 247–250, 1990.

[5] L. Fu. Adaptive signal detection in noisy environments. *The Journal of Neural Network Computing, Spring Issue*, pages 42–50, 1990.

[6] R. Gorman and T. Sejnowski. Analysis of hidden units in a layered network trained to classify sonar targets. *Neural Networks*, 1:75–89, 1988.

[7] R. Gorman and T. Sejnowski. Learned classification of sonar targets using a massively parallel network. In *IEEE Transactions on Acoustics, Speech, and Signal Processing*, volume ASSP-36, pages 1135–1140, 1988.

[8] Dan Hammerstrom. A VLSI architecture for high-performance, low-cost, on-chip learning. In *International Joint Conference on Neural Networks*, volume II, pages 537–544, 1990.

[9] D. O. Hebb. *The Organization of Behavior*. Wiley, New York, 1949.

[10] R. Hecht-Nielsen. Neural network nearest matched filter classificaton of spatio-temporal patterns. *Applied Optics*, 26:1892–1899, 1987.

[11] J. J. Hopfield. Neural networks and physical systems with emergent collective computational abilities. *Proceedings of the National Academy of Science, U.S.A.*, 79:2554–2558, 1982.

[12] J. J. Hopfield. Neurons with graded response have collective computational properties like those of two-state neurons. *Proceedings of the National Academy of Science, U.S.A.*, 81:3088–3092, 1984.

[13] S. Jha, C. Chapman, and T. Durrani. Investigation into neural networks for bearing estimation. In J. Lacoume, A. Chehikean, N. Martyin, and J. Malbos, editors, *Signal Processing IV: Theories and Applications*. Elsevier Science Publishers, London, 1988.

[14] M. S. Tomlinson Jr., D. J. Walker, and M. A. Sivilotti. A digital neural network architecture for VLSI. In *International Joint Conference on Neural Networks*, volume II, pages 545–550, 1990.

[15] A. Khotanzad, J. Lu, and M. Srinath. Target detection using a neural network based passive sonar system. In *International Joint Conference on Neural Networks*, volume I, pages 335–340, 1988.

[16] H. T. Kung, D. A. Pomerleau, G. L. Gusciora, and D. S. Touretzky. How we got 17 million connections per second. In *International Conference on Neural Networks*, volume 2, pages 143–150, 1988.

[17] S. Y. Kung. Parallel Architectures for Artificial Neural Nets. In *International Conf. on Systolic Arrays*, pages 163–174, 1988.

[18] S. Y. Kung and J. N. Hwang. A Unified Systolic Architecture for Artificial Neural Nets. *Journal of Parallel and Distributed Computing*, 6:358–387, 1989.

[19] G. Lev, N. Pippenger, and L. Valiant. A fast parallel algorithm for routing in permutation networks. *IEEE Transactions on Computers*, C-30(2):93–100, Feb. 1981.

[20] R. Lippman and B. Gold. Neural classifiers useful for speech recognition. In *International Conference on Neural Networks*, volume IV, pages 417–426, 1987.

[21] R. P. Lippman. An Introduction to Computing with Neural Nets. *IEEE ASSP Magazine*, pages 4–22, April 1987.

[22] A. Maccato and R. de Figueiredo. A neural network based framework for classification of oceanic acoustic signals. In *Proceedings of Oceans '89, Seattle*, pages 1118–1123, 1990.

[23] W. S. McCulloch and W. H. Pitts. A logical calculus of the ideas immanent in nervous activity. *Bull. Math. Biophys.*, 5:115–133, 1943.

[24] Manavendra Misra and V. K. Prasanna Kumar. Efficient VLSI Implementation of Iterative Solutions to Sparse Linear Systems. In J. McCanny, J. McWhirter, and E. Swartzlander Jr., editors, *Systolic Array Processors*, pages 52–61. Prentice Hall, 1989. Proceedings of the 3rd Int. Conf. on Systolic Arrays.

[25] Manavendra Misra and V. K. Prasanna Kumar. Massive Memory Organizations for Implementing Neural Networks. In *International Conference on Pattern Recognition*, June 1990.

[26] Manavendra Misra and V. K. Prasanna Kumar. Neural network simulation on a Reduced Mesh of Trees organization. In *SPIE/SPSE Symposium on Electronic Imaging*, Feb. 1990.

[27] David Nassimi and Sartaj Sahni. Parallel Algorithms to set up the Benes Permutation Network. *IEEE Transactions on Computers*, C-31(2):148–154, Feb. 1982.

[28] K. W. Przytula and V. K. Prasanna Kumar. Algorithmic mapping of neural networks models on parallel SIMD machines. In *International Conference on Application Specific Array Processing*, 1990.

[29] K. W. Przytula, W-M. Lin, and V. K. Prasanna Kumar. Partitioned implementation of neural networks on mesh connected array processors. In *Workshop on VLSI Signal Processing*, 1990.

[30] C. S. Raghavendra and V. K. Prasanna Kumar. Permutations on ILLIAC-IV Type Networks. *IEEE Transactions on Computers*, C-37(7):622–629, July 1986.

[31] U. Ramacher and J. Beichter. Systolic Architectures for Fast Emulation of Artificial Neural Networks. In J. McCanny, J. McWhirter, and E. Swartzlander Jr., editors, *Systolic Array Processors*, pages 277–286. Prentice Hall, 1989. Proceedings of the 3rd Int. Conf. on Systolic Arrays.

[32] Sanjay Ranka. A distributed implementation of backpropagation. Manuscript, Department of Computer Science, Syracuse University, 1990.

[33] Sanjay Ranka, N. Asokan, R. Shankar, C. K. Mohan, and K. Mehrotra. A neural network simulator on the Connection Machine. In *Fifth IEEE International Symposium on Intelligent Control*, 1990.

[34] R. Rastogi, P. Gupta, and R. Kumeresan. Array signal processing with inter-connected neuron-like elements. In *International Conference on Acoustics, Speech, and Signal Processing*, pages 54.8.1–4, 1987.

[35] D. E. Rumelhart, G. E. Hinton, and R. J. Williams. Learning internal representations by error propagation. In *Parallel Distributed Processing: Exploration in the Microstructure of Cognition*, volume 1, chapter 8, pages 318–362. MIT Press, Cambridge, Massachusetts, 1986.

[36] D. E. Rumelhart, J. L. McClelland, and the PDP Research Group. *Parallel Distributed Processing: Exploration in the Microstructure of Cognition*, volume 1. MIT Press, Cambridge, Massachusetts, 1986.

[37] Soheil Shams and K. W. Przytula. Mapping of neural networks onto programmable parallel machines. In *International Symposium on Circuits and Systems*, May 1990.

[38] P. Simpson. *Artificial Neural Systems: Foundations, Paradigms, Applications and Implementations*. Elmsford Press: Pergamon Press, 1990.

[39] Planning Systems. Sonar classification with neural networks. *NeuralWare Connections*, 1(1), 1989.

[40] S. Tomboulian. Introduction to a system for implementing Neural Net connections on SIMD architectures. Technical Report ICASE No. 88-3, Institute for Computer Applications in Science and Engineering, NASA Langley Research Center, January 1988.

[41] B. W. Wah and L-C. Chu. Efficient mapping of neural networks on multi-computers. In *International Conference on Parallel Processing*, volume I, pages 234–238, 1990.

[42] T. Watanabe, Y. Sugiyama, T. Kondo, and Y. Kitamura. Neural Network Simulation on a Massively Parallel Cellular Array Processor: AAP-2. In *International Joint Conference on Neural Networks*, June 1989.

[43] Stephen S. Wilson. Neural computing on a one dimensional SIMD array. In *International Joint Conference on Artificial Intelligence*, pages 206–211, 1989.

[44] X. Zhang, M. McKenna, J. P. Mesirov, and D. Waltz. An Efficient Implementation of the Backpropagation Algorithm on the Connection Machine CM-2. In *NIPS*, 1989.

INDEX